Productivity, Decluttering & Project Management Mastery: Save Time, Stop Procrastination, Master Planning & Get Organized With 2 Books In 1 - Productivity Hacks & Project Managing For Beginners

Table Of Contents

The Productivity & Decluttering Master Plan: Ninja Mind Hacks, Secrets & Habits To Become Productive & Organized, Stop Procrastinating, Save Time, Master Minimalism & Declutter Your Life, Home & Mind!

Chapter 1

An Introduction to the Productivity and Decluttering Master Plan

One of the most daunting problems we face in the modern world is how to be more productive and how to declutter our lives. This book will study and recommend the most effective methods that we can utilize to achieve the changes we need. I will endeavor to explain techniques that will help you become more productive in a clear and concise manner.

Decluttering is something that most of us need to work on to become more productive. The modern trends in downsizing can demonstrate this. While we may not need to go so far as moving into a tiny home, the current reality is that we accumulate too much of practically everything. We all have room for improvement in our lives, and concentrating on decluttering is a positive step in becoming more productive.

Throughout the rest of this book, I will teach you everything you need to know to become the most productive person in your group. I will suggest strategies that will work both in business and on a personal level, which will lead to a more comfortable and more prosperous lifestyle.

Modern life is competitive; we all need to adjust and consider ways to get an edge on our competitors. Increasing productivity is an obvious way. Old habits can be hard to change, so we need to work on our mindset as we attempt to make positive changes in our lives. The power of positive thinking, as it were.

This guide will include plenty of detail that will help you affect dramatic changes in the way you approach things in your work and home life. Your aim should be to achieve the right balance of productivity in work and high quality of life.

We should have an active and productive day at work and still have time for a relaxing glass of wine and a great meal at the end of the day. None of us wants to work too hard, and there should always be plenty of time for family and friends. Increasing productivity at work is a great way to achieve this.

We need to learn methods which mean we can effectively use our time and effort to get what needs to get done - done! There is no single way to reach maximum efficiency; it is instead a matter of making incremental improvements over time.
We will study the methods used by highly productive people and see what we can learn from them. There are many ideas to consider, and we will decide on the best combination to put into effect. They will be simple to implement, but effective.

I will suggest innovative tactics which will suggest things like Ninja mind hacks and secret habits to make your quest to become more productive more accessible. I will discuss how to be more productive in everything and anything you do and how to stop procrastinating and become a "doer."

Procrastination is the enemy of productivity and something that needs to be overcome to become efficient. There are various methods to discuss and then put into effect. There are certain habits of particularly productive people that we need to consider and analyze to see if we should implement them in our daily lives. There are some habits which we will consider dropping as they have been scientifically proven to hinder productive behavior.

Taking on too much work, a failure to say no, and not being decisive enough when it is time to delegate are perfect examples of this. Each issue does have an effective solution. I have run my own business for the past 15+ years, and it has become a personal quest to increase productivity both on a personal level and for my employees.

There are straightforward steps to be taken to improve team spirit and increase confidence. Sometimes it is as simple as some encouraging words, but even that can have a dramatic effect on productivity. Showing confidence in your team and encouraging them to learn new skills and take more responsibility are essential and practical steps to take.

Keeping up to date with technology and taking advantage of new hardware and apps is a constant theme with truly productive people. We will take a look at how best to follow their example. I will attempt to inspire you to achieve more with the benefit of my years of experience and hope to find the key to your productivity.

The fact is that we are all able to become more efficient and become more productive workers. All of this is applicable for both a billionaire business owner and a staff member with lesser responsibilities.

As I mentioned, there is no magic wand or one-step process. It is natural for us to take on new responsibilities as we gain more experience, so keeping our productivity high is an ongoing process. Time management is a skill we can all work on improving, and it is an evolving process. We will consider the latest methods which can help you become more productive.

I can offer you no "all solving" plan, or software, or a way to plan your which will solve all of your productivity issues. I can, however, offer you a series of methods which will have an overall effect. In a typical workday, there is much to do, whatever your responsibilities are. It is worth giving thought to how we can do more. Working harder is not always the best choice; working smarter is.

If you spend a little time the night before contemplating the day ahead and putting a list of goals to be achieved the next day along with a schedule together, you will increase productivity significantly. Whether it is motivating staff, selling products, building something great, merely answering your phone calls and emails or whatever it may be, we can all improve our efficiency.

Plainly stated, the more efficient we become, the more we get done. Productivity will lead to a higher income and a better lifestyle, so let's find the best solutions together! There are specific tools which I can recommend, which will make you more productive and provide you with tactics to effectively manage your time. There will also be tips on creating an environment in the workplace which stresses cooperation and teamwork to streamline operations.

Stress management is an essential factor to consider, and we will study this and recommend ways to keep your enthusiasm high and not risk burning out. Learning to relax, perhaps learning meditation and spending quality time with your loved ones and friends are all vital elements to remaining stress-free.

Positivity and realizing that all of us make mistakes is essential in understanding the appropriate steps to take to create efficiency. I aim to help you on a personal level and also to suggest techniques which will be useful in your business environment.

The way you plan your workspace is essential. It is vital that you feel comfortable. All of this will help you keep a clear mind, which will allow you to be more productive. Education is vital, and I would encourage you to help the younger members of your family to consider and learn these useful methods for leading a productive life.

Unfortunately, we usually do not learn to properly consider information, implement problem-solving skills, or develop efficiency during our schooling, but it is never too late! The majority of the methods I recommend here can also apply to your personal life, and it is often surprising how much extra time can be found to spend with family and friends by being more efficient.

If you try out and successfully learn the methods that I will teach you here, you will be well on the way to improving your productivity and getting more done.

Failure should never be considered the end. It should instead be a lesson and an opportunity to improve. The most successful business people have all faced setbacks and improved.

Exercise and diet are essential to consider concerning work; we should put together a healthy diet and a regular exercise program if we want to be at our most productive.

A great habit that most productive people often share is getting up early in the morning. It allows them time to get more done when their competitors are asleep. Having this extra time can be the key to being more productive and for many people the saying "early to rise means healthy, wealthy and wise," is applicable. It can often help with mental clarity.
Many of us get too tied up in what is in front of us and fail to seek advice when we should. Highly productive people make a point of listening to others, and it is a desirable habit to pick up.

If at all possible, we should choose a business or job which we love to do. It will make doing the necessary hard work and being productive a pleasure. There are a limited amount of hours in a day, and we need to learn to make sure they all count! A successful life is not only judged on what we achieve in our work lives, but also in our relationships and the imprint we leave on the world.

Time management, finding innovative solutions, and keeping a smile on your face even during the toughest of times are all skills that we will consider together.

Join me in the following chapters, and we will look at all of the methods I have mentioned and learn how to put them all together to be the most productive person you can be.

Chapter 2

How to be More Productive in Everything You Do

There is little doubt that we can all find ways to improve our productivity. Perhaps the best approach is to make incremental steps in everything we do. Small steps lead to becoming more productive.

We all have multiple things to do to keep our lives running correctly. We have all experienced working hard, having to keep our family happy, and having a social life. The good news is humans are uniquely able to balance multiple responsibilities and even consider the best way to streamline what we do. We can function productively, even if it seems we are being overwhelmed with calls upon our time. The key is to keep a clear head, prioritize, and maintain a sense of humor.

All of this is an essential way of keeping in control of our lives and staying productive in a positive fashion. Common sense does go a long way when considering our best way to stay productive. We are all aware of how to efficiently use our time, but to stop feeling "snowed under" by life's responsibilities, we should consider the small changes that we could be making.

I will be looking at various methods to use these small changes to become more productive. These will be based on practicalities to help us improve. Most of us can do better and make our work, home, and social lives more productive. These seemingly insignificant changes can make a huge difference in our overall lives.

Many of us have choices to make about our work lives in regards to our decision-making process and ability to compartmentalize our work day. Old habits die hard. The simple process of "deciding to be more decisive" can have a significant effect on what we can achieve in one day at work. Maybe it is finally time to tell the office gossip that you are busy!

Prioritizing the levels of the importance we place on each task we need to fulfill into simple categories such as immediate, today, later, one day, etc. can be an effective way to improve productivity. It sounds simple; and it is, but we are all vulnerable to being indecisive and wasting precious time on something which does not need to be a priority.

Many highly productive people focus on clearing their minds and making use of time management techniques. We should learn from this. I will give various tips on how to manage your time at work, and many of these will apply to your personal life also. You can choose to use as many or as few of these as you wish, but I am confident that all of my methods will work for you to some degree.

The Workplace

A modern-day reality is that most of us feel we have too much work to do and not sufficient time to get it done. Perhaps we need to consider a few different tactics. It is essential to be realistic; if we take on considerably more than we have the resources to manage, we will get nowhere. There is little wrong with working hard, but we all have limits.

If we take on more than we can handle, our stress levels will rise. In my opinion, the key to productivity is to reach a balance. Do not be afraid to say no to extra work. Take on what you can realistically achieve, and don't be afraid to delegate whenever necessary. We all have limits on what we are capable of doing. The fact that we realize where our limits are is a big step to reaching maximum efficiency.
I am a big proponent of getting enough rest as it enables us to operate at an optimum level. "Early to bed, early to rise" is an expression most of us will remember from childhood, and it contains a lot of wisdom. We can learn from this.

You should make sure you like your surroundings in your workplace. We are all different, so identify where you feel most at ease. Being frustrated and feeling trapped is no way to live, so clarity and honesty with ourselves and others when it comes to what we hope to achieve is essential.

Training in the Workplace

Whichever level you work at, and whether you work for yourself or a huge, multinational company, it is crucial to keep up to date with the latest training. There is no doubt that the workplace is changing as technology develops at a tremendous rate. Old knowledge is outdated quickly. Whether it is signing up for a few online tech classes or taking advantage of your firm's latest training programs, it is vital to stay up to date.

The definition of the work we need to complete is no longer so clear. Many of us may feel that we could never finish it all to a satisfactory level, no matter how much time we have. All of this is where being realistic, deciding what is necessary, and prioritizing what we need to do comes in. Sometimes we must accept that not everything will be complete today. The fact that there is no easy definition of the work we do means we need to be decisive to find the best solution. We need to decide where the line for quality and content needs to be.

It is another reality that communication is becoming an increasingly important skill. Cross-department cooperation and discussion are necessary, and dealing with people is becoming an essential skill. Our jobs are changing, and very few of us spend time strictly working on what we were hired to do. The boundaries of what we will be expected to do are changing.

For example, a humble online writer ten years ago would have just been expected to provide the simple copy, but now Search Engine Optimization and providing links to reference sites are standard obligations. Keeping abreast of the latest innovations and training opportunities will enable us to become more efficient compared to our competitors.

So, we all need to be accepting that our responsibilities are increasingly flexible and consider how to equip ourselves best to deal with that. If we were able to assume our work description would stay the same, it would be far easier to master our work productivity. Unfortunately, this is a rare case these days. There are several reasons for this, and the main one is the trend to change jobs or the company we work for far more often than previous generations. A "job for life" rarely exists anymore.

Many of us work on flexible contracts and even for multiple employers at the same time, so it is understandable that different and evolving skills will become routine. Those of us at a later age, say in our forties or fifties, are more likely to change jobs than in the past, so there is a need to learn new technologies and business methods.

Many companies we work for or deal with are also likely to be more flexible in what they offer to clients, which will require more of us. A local food store is unlikely to supply food to only walk in clients anymore. They are likely to have a website and offer home delivery. So, the responsibility of an employee will change.

A manager of such a business, for example, would now have a much broader clientele to deal with and vastly more responsibilities, so learning new skills would be essential. Time management, responsible decision making, mastering e-commerce skills, and diplomacy with clients are all increasingly important.

Responsibilities seem to have blurred over the years; the best we can do is to stay adequately trained and keep up to date with business trends. There is a significant commitment we need to take to absorb the massive amount of information provided in the internet age. Daily, there seem to be new programs and apps to become familiar with and master. It hardly seems possible to be employable without access to the knowledge the internet provides us. We can assume that all of our competitors are up to date.

For instance, a new marketing campaign without considering social media advertising seems unlikely, but ten years ago, it would never have entered our heads. The internet has changed the world, and we need to adapt and change with it or be left behind in the dust. Technology is close to being able to replace us for many work functions with robots, so we need to stay competitive.

Old Work Habits are Obsolete

We can no longer rely on our education from school or college to get by in the modern work environment. Even a fresh graduate will be behind the times in a few months. In years gone by I remember people using a diary for appointments and considering it the height of being organized.

Then we went through a period of using online appointment setters or devices such as Palm Digital Assistants (PDA's), and the reality is today's business world is more sophisticated. We each seem to have a smartphone, tablet device, or laptop permanently attached to the end of our arm, so appointments or communication can be organized immediately. We need something more adaptable and able to assist us with the speed and adaptability required to be a productive member of the modern workforce. The ways we used to use to manage our time are no longer sufficient. We need to ensure we keep up with modern technology.

There are new methods used, and the day when a simple "to do" list was sufficient will not be seen again. The new reality is we may be dealing with 100's of work-related emails per day, and all of them seem to need an answer or an action immediately. We need to be disciplined and not let this take over. It is essential to set aside a certain period each day to handle work-related emails, or we risk losing track of other vital jobs. For sure, we need useful tools to assist us, but we also need a new way of thinking. It is vital to make immediate decisions and become more efficient.

Business Gurus and Training

It seems impossible to go online without being confronted by some new business advisor promising riches if you follow him or buy his course. I have reviewed many of these business training courses or money-making opportunities, and all I will say is consider all sides when choosing them. That is not to say there is not honest advice from good people out there, but for the majority of these courses, it is purely a money-making venture from which you will gain little.

There comes the point when you have to question why there are so many "Gurus" out there who have the perfect secret to wealth and productivity? Surely, if they were that talented and wise, they would be using their skills to make their own money in business. The sad fact is, most of them don't have much business ability, but they are good at persuading you to pay for their training course and then sell it on their behalf through affiliate marketing.

A lot of this is designed to use you to make them ever-increasing profits. There is a more excellent and lucrative use of your time. They typically encourage thinking of "the big picture," which is not always a good thing. Getting the necessary day-to-day things right is still a key element in business success. While we may think that identifying our values and clarifying goals will help us become better organized and efficient, there is often too much going on in day-to-day business to focus on that.

In my opinion, you need to have as much information as possible, and then you will be your own best guru. The bigger picture will not help you when you have 100 emails to answer. Focusing on "the big picture" is essential, but it will not be much help in our quest to become more productive. If anything, it will confuse the issue and slow us down.

We Need to be in Control to be Productive

If we think back to times when we can honestly say we were at our most productive, it is almost certainly a time that we felt in control of what we were doing. If you are stressed out or worrying about an issue with your boss or staff, you will be likely to overreact to things and achieve less and be ineffective.

We need to find a level of calmness and relaxation to be genuinely useful. There are many ways that we can achieve this, and you will need to find out what works best for you. For some, it might be meditation or simply calmly reflecting on your day. Others may achieve this through a quiet meal with their partners; and others through an evening of drinking with colleagues. The point is to find a way that enables you to have the mindset that you are in control. All of this added to a clear mind will allow us to operate at an improved productivity level.

The trick is, when you feel out of control, overwhelmed, bored, what do you do to get your feeling of control back? Later in this guide, I will run through strategies which have worked for me in the past, and it will be for you to adapt them and discover what works best for you. We are all able to improve.

How to Deal With Yourself

One of the main challenges we face in improving our productivity is how to maintain focus and have the right mindset. We tend to make promises and deals with ourselves, which affect the way we think and impact a positive mindset. We need to change this and find a better way of dealing with things that are on our mind. It seems that the human mind is predisposed to complicate things for itself. We need to think of methods to overcome this and reach a state of clarity.

We can commit with ourselves to do practically anything. It could be large or small. It could be to fire your assistant or to buy a subtler brand of coffee for the office. Whatever these deals with yourself are, it is crucial to act on them right away. If you let them fester, you will be distracted and lose efficiency. You are likely to have made more of these deals with yourself than you realize, and each one will be on your mind at some level. They will be dragging your attention away from where it needs to be. You have to be decisive and handle all of these internal commitments so that you can move on.

There are methods to achieve this. There should be a pledge to recognize and list these deals you make with yourself and keep working through them until they are all solved. If they are playing on your mind, you will never have clarity and be able to perform at your most productive level. A practical method to deal with this is to decide what major deal you have made with yourself, and then work your way down the list. It could start with something significant and run to something seemingly insignificant. The critical part will be to decide to solve each one. In fact, write it down to remind yourself.

These will enable you to feel that you have taken some practical action to overcome these issues and give you a feeling of having found some control. This is important for your mindset. Just running these issues through your thought process and committing to do something about them will be enough to help you become more productive. If you are mindful of problems, you can improve them.

We can Train Ourselves to be More Productive.

It is possible to train ourselves like a martial artist or a boxer does, to be more responsive, quicker thinking, and faster overall. While training the body, and being in good condition is an advantage for most of us, the training here is mental. Having clarity of mind is essential if we wish to be productive. We can learn to think more efficiently and either resolve or tune out distractions in either our work or personal lives. The mind is a stunningly powerful tool, and it can be taught to be disciplined.

The aim is to achieve more by using less time-consuming effort. You can make rules to overcome unnecessary distractions and find a solution to being overburdened in all aspects of our lives. You need to keep a clear mind; and one effective way to achieve that is to manage what you do. There are only a certain number of actions you can undertake in one day. You need to make them count.

You need to consider what you will achieve in a day, with your work, with your personal life, and with your mind. Find a way to manage your actions. You must clarify what actions will allow you to move forward. You may have a work project which seems impossible, but finding the right step to take will solve it. Being organized and aware of the issues you face and having a desire to solve them are a big part of becoming more efficient and productive. Learning to identify issues at the beginning of a project and not leaving then until it is too late is an important skill to utilize. Take action at the beginning!

Clearing Your Mind is Key

To reach the feeling of being in control, you need to find your optimum state of mind. In my opinion, the best way to do that is to clear your mind thoroughly. There is no way to be efficient with a cluttered mind. I always try to empty my mind to be able to focus on tasks at hand. You need to leave your troubles and concerns behind you and focus on what is essential right now. There are always issues to deal with, so I recommend you separate them from your mind and use whatever tools you have at your disposal.

You can list the things that bother you then find an appropriate action. You can delegate; you can find an immediate solution; or you can decide it doesn't matter, just so long as you deal with it. It is essential not to be distracted by incomplete thoughts. All too often, we overburden our minds with tasks or questions we can either solve or disregard.

Many of us find that our minds wander when we should be focusing on a task at hand. It is due to something unresolved preying on our minds. How can we be efficient if that is occurring? All of this can cause stress and limits the amount of time we have to dedicate to anything we wish to achieve.

Too many people have a constant state of stress because they have too many things on their minds, which means they can't be productive, which in turn means there is not enough time. It is all interrelated, and without making demonstrable changes, we will never find a way to improve our efficiency and be at our most productive. We need to find ways to relieve ourselves of this stress.

We usually underestimate how much pressure we are under and only notice when we feel better after solving the underlying problem. This is why we feel like celebrating after finally completing a long and challenging project. It is a similar case here; solve the issues you face, and you will feel in control of your life. In short, we need to free our minds to feel in control, and this will enable us to work efficiently and operate at a maximum level of productivity.

Chapter 3

How to Declutter Your Home, Life, and Mind

One crucial way to boost our productivity is to declutter our lives. This means in all forms - our homes, our lives in general, and our minds.

Let's first consider our home. There are practical reasons why we need to declutter it. We can then live a more ordered and less distracted life. Recent years have seen a trend for people downsizing in their personal lives. The trend toward tiny houses is inspiring. It proves that many of our possessions are not needed to be happy. We really should analyze just how much "stuff" we need to keep. Does the football shirt leftover from college need to be saved? There is an excellent argument to be made that having fewer possessions, means we have less to worry about - so we will be better off.

Decluttering Your Home

There are different reasons why we allow our home to get cluttered and disorganized. It could be the stresses of everyday life, laziness, or something more complex.

Taking care of a family, working hard, and an active social life could all play a part. Often a new baby, or even pet dog, can disrupt things and lead to a cluttered environment. Any changes in circumstance, perhaps a new job or being away from home for a training course, can cause us to lose control of our ordered lives. We need to identify and correct this.

Once things get out of control, a lot of us subconsciously give up and accept the clutter building up around us. It is a symptom of a stressed mind and will adversely affect our productivity levels. The question is, what can we do about it? To be productive, we need to have organized surroundings. A clean and decluttered house is a necessary part of having a clear mind. We can't expect to be living in chaos and to be at our best. We have to stop being disorganized, stop putting things off, and make some decisions. It is time to make some changes!

Whatever your path to clutter has been, you need to make a conscious decision to make a change. Analyze why it has happened and identify what to do about it. The following section of this guide will make recommendations; and you can use it to help make positive changes to your surroundings. We need to get rid of the chaos and find some order.

Does my Home Have an Issue with Clutter?

You would think that it would be easy to tell if you have too much clutter; but sometimes our familiarity with our surroundings means we don't notice it building up until it is overwhelming. The signs are when you start to notice there are papers everywhere. Perhaps there is too much dirty washing or kids' toys.

This can get increasingly worse until you start to find newspapers from a few years ago and to have difficulty finding a place to sit down. It is easy for our minds to find excuses when this happens. You may have trouble finding things that you look for and end up buying things you already have. Does this sound like a person in control? Try to improve your awareness of this. It should bother you.

Before we know it, we get used to this state of calamity, and it becomes the new normal. It can start to feel like there is no way of changing it. Living like this means you are making things unnecessarily difficult for yourself, and there is undoubtedly a better way. You cannot be at your best in such a complex environment. The good news is you can easily decide to do something about it and create something new and improved!

How to Make a Change

Firstly, learn to notice the signs that the clutter in your home is getting out of control. Is the washing up always piled up in the sink and you only ever think of washing it when there are no more plates available to use? Is your bed in a permanently messy state? Is this due to your subconscious saying to you, what's the point of making your bed; you will be sleeping in it again in a few hours? Just how long has it been since you properly ironed your clothes? Or hung them up in the wardrobe properly?

Do you have unopened mail and you read newspapers and magazines from years ago, just because they are close to hand? Is your garage so full of junk that your car has been parked out in the street for months? Are you afraid to let your family and friends see the state of it? Are the kids' toys everywhere, and are they following your approach to leaving things lying around? Lead by example, not by sloppiness!

Do you have many, different half-used soaps and other toiletries in your bathroom? Are the cupboards full of things you will never use? Is everything so cluttered that you are starting to have trouble finding your car keys in the morning?

If all, or even some of, these questions are an issue in your house, it is time to make a positive change. You are living chaotically, and there is no way your mind is at its most productive.

Firstly, recognize that there is a problem and decide to make a change. Write it down and leave it somewhere you will always see it. Keep a list of all of the inconveniences that the state of your house causes. These could include having to wash plates when hungry or never having an ironed outfit ready in the morning. Keeping a record of all of this will emphasize to you that there is a problem and it will serve to keep you determined to do something about it. There is a better and more ordered way of living!

Put a System in Place to Start Decluttering

One step at a time is the key here. You have already made a good start by writing down the inconveniences the state of your home is causing you, and your determination to make positive changes.

Don't be overwhelmed by the task at hand and get into negative thought patterns such as "I will never fix this. It is pointless to try." Also, don't be tempted to do a quick fix by putting all of your junk into a cupboard and pretending it has gone away. This situation needs a proper solution! There is no point by trying   to   fool   yourself.

Don't try to do everything at once and make a start all over the house. Discipline and planning are necessary to regain control. Develop a plan to tackle each part of your home in stages. Why not make a start in the kitchen? A clean area to prepare food and have all of the plates and cutlery in the right place will feel like a positive step. Breaking the project into parts will make it feel like a more straightforward process and encourage you to keep going.

Visualize the result and think of a well-organized, clean, clutter-free home and how good that will feel. It is an essential step to regaining the feeling that you are in control. The fact is you have adapted to your old careless ways, and it has become the norm. You need to retrain your brain and pledge to yourself that once this is solved, you will not go back to your old ways.

Some houses you visit may seem well-kept, clean, and orderly on the surface, but have drawers and cupboards full of junk when you look deeper. Don't let that that be your future! This is a trap to avoid, as we are not interested in appearances, but having a truly decluttered environment. Having something beautiful to show people is not enough. The fact that you tidied up and did a bit of vacuuming is missing the point I am trying to make here. You need to make a real change rather than a cosmetic one.

When you finally do get organized, you should aim to have a purpose for every cabinet and drawer, and not just use it as a storage place for unnecessary junk. Mindfulness is a part of having a clean home. We should plan everything in an organized way. All of this may seem inconsequential, but it is a factor in becoming more efficient. Our mental clarity is vital to our productivity and the environment we live in today. A disorganized and chaotic place will affect you in myriad ways.

The next time you come home, imagine that it is someone else's place. What is your impression of it? Have a look around; open some drawers. Does it seem like an organized place? Is it a real mess? Are some areas, like the kitchen or garage particularly chaotic? What is your overall impression? If it is negative, there is an issue which will need to be worked out. Giving yourself some distance and looking at the state of your home as if you were an outsider is an effective way to start planning for getting organized. All of this may also help you realize that things are not as bad as you had feared and make starting the decluttering process seem possible and realistic.

You need to get a system in place. Establish the main problem areas of your home and make a start. One room first; and commit to getting the task complete within a fixed schedule. Another right way ahead is to develop some day-to-day practices and stick to them. You can try to clean the kitchen every day, including all plates and cutlery. Throw out all of the garbage each evening and make the bed each morning after you get up. Simple things matter!

Make a list of ten things like this to do each day and write them down and make them a habit. Include something unconventional, like brushing your teeth with your other hand. This will act as a reminder to do the less pleasant tasks, and all of this is a useful mental process to start making a change. Subconsciously, you are already becoming more efficient and productive. You can and will find the best system that works for you to declutter your home. And it will feel like a weight off your mind when you do.

Remember that we all can get organized. A written list of new habits to follow will be a big help; and dividing the steps to take will make that process easier. All of these steps we have discussed help us regain control over our environment and get us in a more productive state of mind.

Decluttering Your Home Office is a Way to Improve Your Productivity at Work.

It is a new reality that most of us do at least some of our work at home. Whether that means telecommuting or merely responding to a few emails in the evening, a nicely organized home office is vital. Nobody will work efficiently in a chaotic environment, so leaving the dirty plates from last week's midnight snack and a pile of dirty clothes next to the computer will never be a good thing.

An excellent place to start will be to do a quick fix. Take a few hours to throw out anything unnecessary and give the room a good clean. You should take this opportunity to make a promise to yourself to never let clutter take over again and organize your home office every morning. Write it down and leave it somewhere visible. By managing the clutter and cleaning up every day, it will only take a matter of minutes. You will never have to face hours of throwing out the junk in this room again.

An orderly home office will help give you a feeling of control, which, as we have discussed, is vital to having a clear mind and becoming more efficient and productive. The next step should be considering the best way to organize your desk. Is there something we can do without or something that we should add? We should only have things in our home office that we need and serve a useful function. Consider your setup and the best way to refine it.

Let's look at a few useful points to consider and create a plan to set up an efficient and effective home office. What exactly will space be used for, and are all of your tools up to date? Perhaps it is now time to upgrade your laptop and invest in some new hardware.

Is your workstation and chair a comfortable fit for you? Do you have enough cabinets for your files, and is there an area to store old documents? All of this may sound simple; but it is crucial. Think of small, practical steps to make your life easier while working at home. Make sure your filing system is within easy reach of your desk, and that reference books are close to hand. Make sure there is adequate shelving to stop the build-up of clutter. Invest in some more bookcases if you need more storage space.

Lighting is essential for most people. It affects our mood, and being too dark or too bright is not advisable, so it is worth investing in some good, adjustable overhead lights and lamps. You need to define what the room is for. Is it strictly your workspace, or can your kids use it for doing their homework? If there are other uses sometimes, like watching TV, give each activity its own space.

These are all simple methods that will make working in your home office a happier experience and help you become more productive.

Effective Methods to Declutter our Life and Mind

Decluttering your home is a good start, but we need to consider the best ways to do something similar with our life and mind. Earlier on, we have discussed that achieving a clear mind and taking out unnecessary distractions is the way to become more productive.

The fact is, many people have a cluttered mind, which in many ways is worse than a cluttered home or workspace. There are a few straightforward ways we can declutter and reach the clarity we need. Too much clutter in our mind will make us restless and make it difficult to focus. It will cause us to be distracted and make it difficult to get things done. Promise yourself that you will change it.

Some things that we can refer to as "Clutter of the mind" are overthinking things, going over past experiences, again and again, having worries about the future and feeling overburdened. In a way, our mind is a little bit like a hard drive on a computer, and there are ways to delete unnecessary clutter and create some space. Give it a regular checkup in our mind.

Let's consider some ways to declutter our mind so that we can have more clarity and achieve more. We have already discussed ways to declutter our physical space at home and work. It is essential to sort this out as excessive stimuli will impede our ability to be productive. We can declutter our lives in other ways. Perhaps we can look at our social lives, and maybe we can miss the occasional drinking session with friends or game of golf.

Write down the steps you decide to take, either in a journal or download an app. All of this helps to store information outside of the mind and free up some space. We can add appointments, ideas, etc. Keeping a general journal is an excellent way to expand on the previous point, and a way to keep track of our thoughts. It will allow us to note down our concerns, which act as a distraction to getting things done - and ultimately, being productive.

You can keep a written record of the following:

- Goals which are essential to you, and how to achieve them.
- Things which affect your confidence and energy, such as worries about a relationship
- General things which prey on your mind and worry you.
- Everyday tasks that will help your life stay more ordered.

Keeping a journal is also an effective way to discipline our minds, by taking a set period per day to write and reflect on what our day was about and what needs to be focused on tomorrow.

You should let go of the past. Let it go and don't let it bother you. Whatever has happened, put it down to experience and think of the positive. Plans for revenge or changing the past don't work out well.

Multi-tasking is not always practical. There is a time and also a place for everything. Devote a set period of time to each task. Let your mind be free and ignore distractions that come into your mind. Choose to be more decisive. Most of us feel overburdened at work and often with our home lives. We need to decide what is most important and deal with those issues one by one.

Many of us suffer from receiving too much information, whether it is from newspapers, surfing the internet, social media, or watching too much TV. Again, this will lead to a lack of clear thinking. To avoid being preoccupied with all of his information, we should limit what we take in. We should set a time limit on our time spent online and choose the content we follow carefully. Make a conscious effort only to read or watch things which are of benefit to you. Try to cut out the nonsense and only take in relevant information and ignore the rest.

We should be more decisive. In the same way that we should decide what to do with the 100's of emails we receive, we need to do similar with the information our brains, or it will be overcrowded. A cluttered mind will not function properly. Your decision-making process is essential. And don't procrastinate. Decide what to do with the information you have. Either disregard it or take action on it.

The small things in life can take up too much time and energy. For mundane tasks such as breakfast, what to wear, and what to take to work for lunch, make a policy and stick to it. There is no need to spend time each day considering the small things over and over. Get into a routine and get the little everyday things off your mind by doing them automatically.

Meditation may be a useful practice to help clear and discipline the mind, whether it is a formal process such as a Buddhist may use or only taking some time to do some deep reflection. Basically, meditating will allow you to focus on one thing at a time - for example, breathing - and clear your mind. One thing at a time is always a good policy to adopt. It is almost like taking your mind to the dry cleaners and getting all of the wasteful clutter washed away. Most of us need this. Meditation is undoubtedly an effective way to get rid of unnecessary thoughts, regain focus, and have a balanced feeling which will help us become more productive.

We need to learn to prioritize every part of our lives to feel happy and have a clear mind. An endless list of things to do will clutter the brain and lead to being unproductive. We should know what our top priorities are and focus most of our brain power on doing those things. There is no point having an endless to-do list running through our minds.

We have seen there are various ways to declutter multiple parts of our lives. We need to choose which are the most applicable to us and decide on the best methods to deal with it. Clutter in our minds leads to a build-up of random thoughts and information in our inside worlds. It stops us being able to think clearly and affects our ability to focus on what is essential.

Decluttering our mind is one of the most powerful things we can do if we wish to be more focused. Seriously consider working on this. Following the methods mentioned above will be an ideal way to declutter your life and will enable you to feel mentally clear. Declutter your home, life, and mind, and you will be well on the way to achieving an optimum level of productivity in your life. You will notice improvements at work and in your life at home.

Chapter 4

10 Time-Saving Secrets to Effortlessly Beat Procrastination and Become More Organized

Procrastination is something even the most disciplined and effective amongst us will face from time to time. It is the enemy of productivity. For some of us, being under pressure as we have left things to the last minute is a worrying thing to dread and will lead to panic and a substandard quality in our work. For others of us, overthinking things and leaving everything until the deadline is almost here is a way of life and a situation to thrive. We are all different and work best under different conditions.

However, for many of us, procrastinating will make our home and work lives suffer. And when that starts to occur, it is time to decide to do something positive about it.

I will now run through a few proven methods to help you with your time management, reduce procrastinating to get more done, and reduce the need to racing against the clock.

Fix a Deadline, and Then Follow it.

One of the best ways to overcome an issue with procrastinating is to set a fixed deadline and commit to it by letting other people know about it. For example, you may need to complete a sales report and not feel into doing it. However, if you tell everyone there is a meeting to discuss it on Monday morning, you won't have time to procrastinate.

It is human nature not to want to let people down and have to rearrange everything, so you will find your mind keen to get on with it. Even if this is a self-imposed deadline, you can treat it as if your boss created it and honor it in the same way. Nobody wants to let the boss down, and your mind indeed won't.

This method works well for me, especially if there is something that needs to get done, and I am struggling to get motivated. Make an appointment to discuss your project and follow it.

Start With a Small Step

We all suffer from times when we don't feel motivated, or when we don't fancy doing what we need to get done. Let's imagine it is a report which needs to be complete by the weekend and you can't seem to get motivated. Something which always helps me in this situation is to take a small step. It may be as simple as noting down the headlines and titles. It will get your subconscious mind thinking about the subject. You may have then given it more thought than you have realized.

I often try to do his last thing at night if I know I have to write the next day. More often than not, I end up making a decent start and getting quite a lot done. All of this will make a massive difference in your mindset. Even the smallest step means you have made a start and makes the overall process seem more achievable.

It will have the effect of disciplining your mind to get used to the activity each day, and you will get used to the idea that it is not so bad. Hopefully, the endorphins will kick in, and soon you will be enjoying the process, and an hour per day will start to fly past. Taking even the smallest steps toward your goal is always a good start, and it will help you fix any issues you may have with getting motivated.

Just taking a step will allow you to overcome any mental block you may have, stop procrastinating, and become better organized.
Put up a Do Not Disturb Sign

Issues with procrastinating can often stop by exercising self-discipline. Lock yourself away, let everyone know not to disturb you, turn off your phone, and set a time for your project's deadline. You need to write down your intentions, when it needs to be complete by, and turn off any distractions which will tempt you to get sidetracked.

If you regularly do this, it will start to be like muscle memory. So, set the scene to do what needs to be done and create the environment to be able to focus efficiently. It is a beautiful way to increase your discipline, overcome procrastination, and become a more organized person. Give it a try and see for yourself.

Don't be so Hard on Yourself.

Many of us tend to feel bad about procrastinating and our lack of ability to get motivated to get things done. We should, too. It impedes our ability to be productive. However, we should understand that we are human and stop being so hard on ourselves. We are not bad-intentioned, not all wrong; we are just human beings.

Instead, we should focus on the positive and concentrate on just getting a little more done each time and getting closer to our objective. Overcoming Procrastination and becoming more organized can be a gradual process, and a step by step approach is realistic.

Accentuate the Positive

It would be best if you always considered the positive outcome and reward that will come with it when you are contemplating doing something.

Whether it is picking up your son's bike from the repair shop or making a phone call to complete a sale for work, most things come with a reward. In these cases, a happy child or boss. It is an effective way for you to overcome a tendency to procrastinate and focus on a positive outcome. The reward is usually worth the effort.

Most of us feel happy after achieving something, and there is often a satisfaction in that, no matter how small it may seem. Motivation comes in many different forms.

We Need to Understand Why We Procrastinate

There is no need to go so far to pay for therapy, but doing some amateur scientific or detecting work is useful to help us understand why we face issues with procrastinating and getting organized. We can start by noting down our thoughts, moods, and patterns of behavior when we feel we are procrastinating.

Some of us may be perfectionists, whether we realize it or not, and the need to get things right can cause us doubts, which lead us to delay things. If we can better understand our pattern of behaviors, we can reduce our doubts and anxieties about this and gain positive thoughts about doing what we set out to achieve.

Understanding ourselves and why we do things will lead to more exceptional organization and increased productivity. The key is being objective when we consider why we do the things we do.

Reward Yourself When You Complete a Necessary Task

One effective way to help overcome procrastination is to write a list of things that need to be done and separate them into things you feel happy to do and those that you can't get motivated to start. Please start with the least appealing task and assign yourself a reward for when you complete it. It could be something simple like a chocolate bar, a nice glass of wine, or an hour of your favorite TV show.

Next, do one of the more delightful tasks, and after that, alternate. It will make it mentally more natural to work through all of the functions. The brain does enjoy getting rewarded and will respond to this method.

It is an excellent way to get more organized and start to achieve more. If increased productivity is the ultimate aim, little tricks like this that we can play on ourselves are an effective way to make it.

Discuss Your Schedule with a Partner

Choosing a partner, possibly a work colleague, whom you will feel accountable to, is an excellent way to overcome procrastination issues. It can work in favor of you both as you can both encourage each other and feel responsible for your partner to get things done on time.

The mind is a strange thing. And it hates to let other people down, so making that commitment can be a truly effective way to get properly organized. It can be employed in other parts of our lives aside from the workplace as well. If you commit to saying you will lose weight by going jogging every day and discuss it with a partner, it will help with motivation. You will avoid thinking about getting up and going for a run in the morning if you have made a commitment and promised your partner. No more procrastinating!

Many of us know we need to achieve something and we keep delaying it as we do not commit to following it for anybody else. It is always too easy to make excuses for ourselves. Discuss the schedule with a partner and get their help to help make you follow it and make it clear that you will do the same for them.

It is essential to have the right partner whom you will feel accountable to, and he or she will be able to assist you to overcome procrastination and become more productive.

Have Some Set Daily Tasks

There will always be times when you will not feel like doing something. An excellent way to overcome this is to set some failsafe daily tasks, a few things that you commit to doing, no matter what happens or how you feel; something that you can't negotiate about with yourself.

It could be anything which is strictly necessary. For example, it could be a report on the sales achieved by your team and an update of their expense reports while on the road. If you commit to completing both of these tasks at the end of each day, it will become a habit that you will not break as you have already made a firm commitment to yourself. It will become second nature and something you will never need to procrastinate. It will lead to an improvement in your work practices.

Have a Separate Time for Procrastination

There is nothing to stop you from setting aside a period per day to sit and procrastinate about things. However, you should schedule it, and stick to it. Some of us quite enjoy the process of procrastinating. So, if that is the case, we could use it as a set period per day as some others use for meditation. Ideally, you can schedule this during a period that you will not be working. It could be while you are washing the plates or waiting for your child's soccer practice to finish.

It would be best if you tried to be disciplined about it and set a set period every day so that it does not interfere with your work time. Marking this time in your schedule may eliminate many of the adverse effects associated with procrastinating and allow you to have a clear and refreshed mind when you start to do your daily tasks.

To conclude, overcoming procrastination is best done by using disciplines and methods like these. It would be best if you committed to do or change things and stick to it. Use these ten methods to become more organized and stop letting procrastinating being such a negative in your life.

Being more productive is an aim that we should all strive for, and it is entirely possible if we implement a new way of doing things and a new mindset.

Chapter 5

What are the Little Known Habits of Highly Productive People?

There is a lot to be learned from the habits of others, and I would like to take a look at some of the practices adopted by highly productive people. There is a lot to be learned from people who are elite in their field, whether it is business, sports, or the arts. They seem to have a thirst for success and endless energy and spirit. How do they achieve it?

The first thing that becomes apparent is that we should aim to work smarter rather than working for all hours under the sun. Increasing our efficiency is a way to increase productivity.

Let's look at some examples:

You Need to Eliminate Distractions

It is a golden rule for particularly productive individuals. They recognize that this will slow them down, and they will seek out ways to minimize the issue.

I used to have a cleaner in my office, who was unbelievably chatty. The staff member was a nice woman whom I didn't wish to offend, but every time I saw her, she would take up fifteen minutes of my time. Although it was a hard thing to do, I decided I needed to explain that I was busy, and I only had time to chat after work. I blamed it all on my boss!

If your workspace is too loud and you find it a distraction, buy yourself a good headset and put some inspiring music on your phone to listen to while you work. Don't open websites on your computer that are not relevant to the task at hand. There is time to read about sport at home later! By being aware of the distractions you face and making the decision to minimize them, you will increase the amount of work you get done dramatically and become a far more efficient worker.

Similarly, in your home life, forgoing an extra glass of wine in the evening may allow you to get your house cleaned without distraction. You can always have it as a reward later. A minimal amount of disciplining ourselves concerning stopping distractions will lead to greater efficiency and lead to us becoming more productive. It is worth the effort.

Multitasking Will Slow You Down

People who practice multitasking may think that they are achieving a lot, but it is a proven way to be less effective. It slows down our mental process and makes us less efficient. Going from job to job is not an effective way of working. Concentrating on one task over an extended period is the best way to get things done. It allows for deep concentration and effective results.

Imagine a rock star struggling with writing a new album. He is likely to be better off focusing on finishing the song at hand than jumping from one to the other. Discipline and being determined to finish what is in front of us is the best way to be productive. Multitasking is the enemy in his instance.

Practice eliminating multitasking at work and during your home life, and you will notice an improvement in your productivity. Do it today, and don't procrastinate!

Don't Always Focus on Your Emails.

Receiving 100's of emails every day seems to be a current reality. You would be amazed at how much time per day you can lose by leaving your inbox open. It is human nature to want to deal with things as we become aware of them, so genuinely productive people limit the times that they read their inbox.

It varies from person to person, but I like to spend an hour when I first arrive at my work desk, an hour after lunch, and an hour at the end of the day contemplating and answering emails. Being selective of when I deal with it means I get far more done. It just takes a bit of discipline.

If you have someone (perhaps your boss or an important client) that needs to be dealt with urgently, you can always set up an alternative email which will alert you, but overall spend time on other work. It is a particularly effective method used by productive people, and you should try to implement it into your work day. It will make a big difference.

Similarly, it would be best if you disciplined yourself at home to not be a slave to answering emails. Your family deserves quality time with you, so fix a time to check any critical emails. A half hour at night is perhaps the best way to do this. Then you can sleep with the knowledge that you can start the next day with everything up to date.

Taking Small Steps is Key

Highly efficient people will always approach a large project by making it seem more achievable by dividing it up. A massive job may seem overwhelming and make it seem challenging to start. If you make a start and go step by step, you will make good progress.

Make a list of achievable steps to take, and the project will start to feel more manageable. It will motivate you to keep going and know that you are making progress. It would be best if you overcome the feeling that the job is too much to handle and resist the temptation to procrastinate. If I am working on a book project, I like to start with an accessible introduction and then choose a chapter which I am most familiar with the subject and work on that.

We all have our methods to enable us to complete large projects, but making a start and working through it step by step is always a good practice to have. Psychologically, seeing the actual process will naturally inspire us to keep going and make completing a task seem possible. Step by step is an excellent method to help yourself stay motivated.

Implementing this method into our work day will help us become more efficient and get our projects completed more quickly. Try it and see.

Don't be Afraid to Make Mistakes.

The most productive people don't have a fear of failure. When things go wrong, it is the time to use it as an experience to learn from and to encourage us to improve.

There is no need to over-think what has gone wrong. The best course of action is to redouble your efforts and get productive again. There is nothing to gain by being inactive and worrying about making another mistake. People are human, and we need to realize that mistakes happen to the best of us. Don't get down about it.

If you wish to become truly productive in the jobs you do, you should embrace the element of risk. It is an essential way to develop your skills. Accept it as a part of what you are doing. Many successful and wealthy business people have experienced failure and gone broke at some point. Their ability to keep going and improve is what makes them a success.

We should employ this principle to our work, even if the tasks we are seeking to complete are relatively humble. Keep going, and we can always get better! Don't use failure as an excuse to procrastinate and instead use it as a motivational tool to make improvements. It is an essential method to becoming more productive and efficient.

Start the Day by Doing the Most Complicated Task First.

Some people refer to this as "Eat the Frog." It relates to starting the day by getting the most unpleasant task out of the way. It is an idea developed by Brian Tracy, who is a motivational expert and first espoused in his book "Eat That Frog."

The theory is that by getting the worst thing over and done with, the rest of the day will seem more straightforward by comparison. It also has the effect of freeing you from worrying about it all day and putting it off to later in the week. Perhaps your most hated task is writing a long, boring expense report for your teams' sales calls? Try doing that first and getting it out of the way.

So, if you want to increase your productivity, you should make a habit of "eating the frog" first thing in the morning. It will make your day more manageable, and you will work more efficiently. Making a habit of all of these methods together will lead to you becoming a more disciplined and productive worker.

Perhaps on the weekend, the most dreaded task is bathing the dog. Start the weekend by ordering Fido into the bathtub! Everything else will be easy.

Get into the Habit of Getting Up Early

There are only so many hours in a day, and the most efficient people around get an advantage by being up and alert before everyone else. Having some alone time in the morning and being able to contemplate and prepare for the day is one of my favorite things. It helps me clear my mind for the day ahead. Many people find that they feel at their most effective in the morning, so I like to take this opportunity to plan my day, work out what needs to be done, and make a start on a few creative ideas.

If we wish to operate at our most productive, most of us will benefit from some exercise so that we can stay in shape and feel mentally alert. Early in the morning is the perfect time to get it done. You can also use this extra time to do something beautiful, which will be mentally rewarding. Perhaps make your kids a lovely packed lunch to take to school (and make a little extra for yourself).

If you are a late riser or are always struggling to get into the office on time, try getting up earlier. You may be surprised that the extra time you have makes you feel more in control and energizes the spirit. It is an effective way of adding to your productivity and is certainly worth adding to your list of changes to make. Start by setting the alarm a bit earlier each day.

The Things we Eat Have an Effect on Our Productivity.

The most productive people around are aware of the importance of eating regularly and using a healthy and balanced diet. A healthy body equals a healthy mind. It is an accepted fact from the World Health Organization that eating the right ingredients can help boost our brain power and concentration.

Maintaining a healthy balance in our diet and trying to incorporate as many superfoods as possible can make the difference between feeling exhausted and feeling like we can conquer the world. I like to eat a big, healthy breakfast to set me up for the day and prepare a nutritious, packed lunch. It helps eliminate the temptation to order a hamburger and fries or a pizza.

Try to eliminate junk food and sugary drinks, which will make you feel like you will crash, and take the time to prepare something healthy and nutritious. If you get into the habit, it is quite quick to do. You should always follow regular meal times and never skip meals. The body likes regularity. Even if you are busy, skipping meals is counter-productive as missing a meal will lower your productivity. So, taking a 15-minute break and enjoying something to eat is a wise investment of your time.

Don't underestimate the benefits of a healthy diet. Making it a part of your day will make you feel better, and it will help your productivity in the long run.

Rewards for Completing a Task are Important

Productive people tend to make a pact with themselves that completing a task deserves something beautiful as a reward. It is an effective method to become more productive. Psychologically, this can be important, and it will encourage us to keep on going when things get boring or tough. Getting through the tough times is vital if we wish to work at our jobs well, so everything helps.

It is possible to train the brain to become more efficient through teaching it. A reward upon completing a task will subconsciously make us want to get onto and achieve the next job and help us maintain a positive outlook. I have a sweet tooth, so something as simple as keeping a jar of sweet candy on my desk works for me. I reward myself with some whenever I complete a project.

Celebrating as a group with your workmates as a reward for completing a task is also an effective way of becoming more productive and has the bonus of helping team spirit. It is an essential method that productive people use, and you should add it to your day. The reward could be anything, a chocolate bar, or a few drinks after work, whatever motivates you.

Exercise is Important

Did you ever hear the expression, "a healthy body equals a healthy mind"? Taking the time to exercise regularly is something most highly productive people practice. A small investment for a gym membership or an even cheaper jogging habit can work wonders for your productivity.

I like swimming, and going to a health club on the way to work to do my laps makes me feel invigorated and alert for the day. It certainly makes me fitter and helps me stay in good shape. Whatever method of exercise you choose, it should be something you enjoy, and it is crucial it doesn't feel like a chore. You can ask the family to join you so you have another activity to do together.

There is scientific evidence of the benefits of exercise on the mind. It can help with your memory and increase your attention span. All of these are essential factors in making you more productive. Just 30 minutes per day of cardiovascular exercise per day can make a tremendous difference in your life and overall health, so those of us aiming to be more productive should add this to our lives.

Planning the Night Before is Key

Highly productive people understand that worrying is counterproductive. It is far better to plan everything and then get a good night's rest. Poor planning can lead to a stressful day ahead and can give you a feeling of losing control and that chaos is taking over.

It would be best if you considered taking a few minutes before you sleep to plan your next day. It can make all the difference in making your next day a productive one. I like to focus on what I need to get done the next morning and write it down. Even if I have a hectic day ahead, I feel this helps me not to worry and sleep well. I like to wake up in the morning feeling in control of my day ahead and knowing what tasks I will do when I start my workday.

Pre-planning is a vital part of working efficiently and becoming more productive. I recommend that you add this method to your schedule. I am sure that you will notice a difference in your work day.

Learn to Delegate

The most highly productive people realize that it is impossible to do everything. Working hard is certainly a trait that is necessary to become successful, but learning to delegate is equally essential. Even if you have the mindset that success must come at all costs, we are all limited by the number of hours we can work in a week until our productivity starts to drop.

A 35 or 40 hour standard work week is acceptable for most non-disabled workers, but those who push themselves to work much longer hours to achieve success. To become efficient, we need to be able to judge when it is time to delegate. There is likely someone on your team somewhere who is more suited to the task.

I mostly write books for a living, and when I feel inspired, I can write for many hours, but if I go on for too long I lose focus, and I have to spend too much time rewriting because my standard starts to go down. In my case, I find it hard to critique my work, so I write what I think is right and then delegate my work to an editor to review what I have written. It helps to share the workload and ensure quality control.

Whatever your work or business is, you should realize that nobody can do everything and you should seek ways to delegate some parts of your work. Collaboration is a positive thing that allows a different person to give another perspective. We all have different approaches, and we always find ways to learn from each other. If you delegate smartly, you can focus on the overall job and demonstrate the confidence that you have in your team and improves their skill levels. It helps to build a pleasant atmosphere and spirit.

Learning to delegate is an important skill to use if you aspire to be more productive and build an efficient team. Consider what changes you can practically make.

Take Advantage of Technology

It was not all that long ago that we were using a typewriter for reports and a log book for the accounts. The productive people amongst us learn to adapt and take advantage of the latest technology. There are so many apps available these days, which can help us become more productive, that it is vital to take advantage of them. Make a habit of reading about the latest ones.

Some of the best available are Dropbox, Productive Habit Tracker, Trello, and Hours Time Tracking. However, whatever business you are involved in, you will find something of use. As a writer, I was amazed when I first found an app called Grammarly. It can check all of my punctuation and grammar, and even improve the flow and readability of what I write. It has been a Godsend.

We would be foolish if we didn't take advantage of these apps to make our lives easier and more productive. They are there to make our lives easier. It would be best if you investigated which apps will be most helpful to you and start building up what you use. It will give you a boost in productivity and an advantage over your less tech-savvy competitors.

The genuinely productive people amongst us are always looking to get an edge over everyone, and one right way is being knowledgeable about and successfully utilizing technology.

Sometimes You Need to Say "No."

It is human nature to try and do everything ourselves, but similarly to delegating, we need to learn when to say no sometimes. The highly productive person realizes that we are incapable of doing everything, and we need to concentrate on our strong points and what is most important.

When I first started as a writer, it was my instinct to say yes to every job which came along, with the attitude that I would get it done somehow. That is just not a smart approach to working. If you take on too much work, the quality will suffer, so we need to find a polite way to say "no" and focus on the job at hand. One of the most effective ways to say no is to encourage the person who brings the work to you to complete it. You may well be able to advise and guide them to do a good job.

Your time is limited, and learning to say "no" is an important skill to be used if you wish to become more productive. Be selective in what work you take on, and you will see the benefits.

You Need to Set Clearly Defined Goals to Achieve

Truly productive people need to have a clear vision of what they are trying to achieve. You will need to set goals for what you wish to accomplish and within what time frame. You need to avoid being "busy" if you have not clearly defined what you are trying to do.

When I start writing a book, I decide on how many sections and then chapters I wish my writing to contain and then set a realistic timetable. It gives me a goal. Perhaps it is a rough draft of one chapter per day or similar, but the point is I have a goal to achieve and if I do that, I judge the day to he been successful.

Every step you take should be a step closer toward achieving your goal. If you can keep sight of this, you will be taking a giant step toward becoming more productive. It is an important point, so focus on setting proper goals and follow what you decide. It is an effective way for you to increase your efficiency.

Your Work Surroundings are Important

Different work environments work for different people. We all thrive in a space that we feel most comfortable. Think about what it is that you like about your surroundings. The highly productive individual will recognize the value of being comfortable in their work surroundings and the effect it has on productivity.

For some, it can be a conventional office, their bedroom, or even the park. The author J. K. Rowling famously wrote the first Harry Potter book in her local coffee shop. Wherever you feel comfortable is a good place to work. If you prefer to work from home, experience has taught me, you should set a designated work area where you can concentrate and lock yourself away from everyday life.

It is essential to realize that, if you feel comfortable and happy, your productivity and efficiency is likely to increase. Make a point of finding the work location which is best for you.

You Need to Know All of the Shortcuts

Whether you are at home or work, there is no shame in using shortcuts to get the task at hand completed as soon as possible. Most highly productive people practice this. Your time is precious, so if you can get a task completed quicker without compromising the quality, why not do it? Use any useful shortcuts which are available.

An example of this is that in my job, I often need to do much research, so unless it is something of vital importance, I speed read. I can get the essential information in a fraction of the time. I used an app called Spritz Reader to learn speed reading, which has proven to be useful and saves me much time. Deciding to use it has been a wise investment in time and money. Taking the easy route is the smart thing to do in many instances. You should consider which shortcuts you can benefit from and use them!

Learn to Use Automation

It is another useful practice used by highly productive people and can be used in many instances to save much time and be more efficient. It can be a simple message sent by auto-reply to an incoming email informing a client that you will be out of the office for a while or something much more complicated like auto-replies for sales.

Automation is beneficial if you are going on holiday, but can be used regularly and save your business a lot of time and money. I often use automation to handle my email marketing. My business is extremely competitive, and it is vital to answer client inquiries quickly. There are times that I am busy writing, so it is not always possible to handle that myself. So, I found a compelling alternative way. There are remarkably effective programs available now which can answer with high accuracy, often with greater clarity than if I do it myself.

Similarly, I often automate my social media posts. It is a Godsend for me and saves me so much time. There seem to be endless new possibilities with automation. I recently looked at a cloud-based help desk, which appears to be promising. The automated services offered will keep growing, and I am genuinely interested to see what is next. Automated invoicing and lead building apps are now possible, and I am excited to explore that further.

I recommend you look into which facets of your business you can automate. You will be pleased that you did and will be impressed by how much your productivity will increase. Often, the highly productive people we are discussing are that way for a reason. Their methods work! Let's strive to learn what we can from them!

You Need to Identify Tasks Which are Not Important.

You need to be able to decide what you should give priority. What you choose not to do is an essential part of not being productive. It is a vital skill that you need to learn in the battle to become more efficient. What can you ignore or leave for a later date? Perhaps some administration makes no difference?

There is no point in doing something just out of habit. What is the point? Don't be afraid to change things that waste time. We can all find ways to improve our efficiency. I used to meet with my editor every Monday, which entailed traveling through traffic. Canceling it and communicating via more email made little difference but allowed me more time to be productive.

Recognizing non-essential parts of your work day and eliminating them is an effective way of increasing efficiency and becoming more productive. Analyze what is unimportant in your workday and set a plan for eliminating it. You will save more time than you expect; and time is money.

Music Can be Important.

Highly productive people make good use of music in the workplace. It can be an effective way to improve your mood or just background noise to help you get through a tedious task. I love listening to quiet classical music when I write. It calms my mind, and I believe it helps me focus and be more productive and get more things done.

However, one of my least favorite things to do is typing up invoices. I find this dull compared to writing, so I might well be playing the Clash or the Sex Pistols or something else aggressive to help me get through it. Whatever you think might be useful is worth a try. As you are listening through a headset, you can't bother anyone, and there is no way for them to criticize your taste in music!

There are also some ambient sound apps which can help you concentrate if you are in a loud environment. Search for the latest apps available. Sound can be so crucial in our lives. I went through a terrible period of not being able to sleep properly. Something was bothering my subconscious. I found downloading and listening to white noise while I went to sleep solved the problem for me.

I believe music is an underrated tool in our quest to become more efficient. I recommend you experiment with what types of music will work for you. It certainly helps me to focus and get more done. Anything which helps us with productivity should be looked at seriously.

You Need to Know When to Rest Your Eyes and to Take a Break

New ways of working means that many of us spend extended periods staring at a screen, and eye strain is a real issue to cause concern. When I started as a writer, I usually used a notebook to write, but these days I often spend up to 12 hours in front of my computer. I don't think the human mind can efficiently deal with that.

It is an excellent practice to take a little 5 or 10-minute break every hour. Just take a short walk to stretch your legs or make yourself a hot drink.
If we don't do this, it is too easy to get mentally fatigued and can lead to headaches or other ailments. The mind needs to have a regular break to relax and focus on something different.

Similarly, you need to know when to do something different for a while. Many highly productive people seem like superheroes sometimes with how hard they work, but even they take a break. Going to the park for some fresh air or taking a quick nap under your desk is a great way to feel refreshed and be able to operate at optimum productivity again.

Take the time to take a break from time to time. Working non-stop is counterproductive and will lead to cloudy headedness and a lack of efficiency.

Having a Tidy Desk is Important

We should adopt the slogan, "tidy desk, tidy mind." It has some basis. How can we work if we are in chaos? Try new methods of avoiding clutter by getting as much done online as possible. Do you need a paper copy for the expense reports? If you do have a job that demands you use lots of papers, take a vow to improve your filing system and keep to it. It will make a tremendous difference in how you feel, and hence, the quality of your work.

An organized filing cabinet within easy reach of your desk may seem an obvious thing to have, but it is far too easy to overlook. Some of us seem destined to be surrounded by clutter. If you do accumulate clutter during the day, ensure that you clear and clean your desk every evening before you finish work for the day.

Try writing a note to yourself saying that this needs to be done before you finish work and leave it in an easy to see location. Making cleaning up each day a regular habit will allow you to start the next day with a clear mind. Productive people work best in an orderly environment.

You Need to Love Your Job

The last method I will recommend to you is to find something to do that you enjoy. It will improve your life in every way. If you genuinely love your job or business, it is so much easier to be productive. A happy state of mind will always help you to be at your best. All of the advice I have given you so far is unlikely to be useful if you are unhappy in what you do. It is time to do something about it!

Motivating yourself to be productive is so important as the alternative is lethargy, and it is so much harder to do if you are only working on getting by or for the money you make doing it. If you find yourself hating what you do, it is time to reflect on the best way to look at the changes you should make.

Can you make changes in your current job? Perhaps a chat with your boss or a work colleague about changing your responsibilities could help? Possibly being transferred to a different department or finally demanding a promotion would make a difference? If you feel none of these would make you happy, you really should consider a career change.

There are some brilliant adult education programs available these days, and it is certainly possible to combine night classes with a full-time job. I used to run a swimming pool supply business, which was a good business financially. But eventually, I became stale and hated going into the office in the morning. So, I had an honest look at my work life and decided to make some changes. In my case, writing had always been a passion.

I decided to enroll in some writing classes and started taking on some small jobs. It was the best thing I ever did! Now, I feel so much happier and have a degree of flexibility in my life that seemed impossible before. I am indeed a far more productive person since I decided to make that change. I encourage you to look at other options if you are unhappy with what you do. Follow your passion and love what you do for a job. It will make a tremendous difference in how you feel. Life is short, so why waste it in something you don't enjoy doing?

Most truly productive people do what they love. Whether it is running your own business, writing, playing a sport, or designing something - find what you love. We have looked at various practices of highly productive people. They all have their merits. I would like you to consider all of these methods to become more productive and decide which will help you the most.

If you implement these practices, I am certain that you will see an increase in your productivity. Give it a try.

Chapter 6

The Secret Habits and Ninja Mind Hacks to Become the Most Organized Person You Know

In this next section, I will reveal some secret ways to form life-changing habits, which will help you become super organized. There will also be some useful "Ninja Mind Hacks" discussed.

A Few Ninja Mind Hacks to Consider

There is one particular Ninja Mind Hack, which I believe works well, which is always to establish the end of your day. No matter how much work you have to do on each particular day, always have a definitive ending to it. This will enable your mind to focus and be alert to the fact that there is a limited period to get what you need to do completed.

It will enable you to get more done in the limited time that you have as your subconscious mind will be aware of the limits placed upon it and strive to get the maximum possible done. A definitive ending to your day will also help you stop procrastinating and ensure that you will not overwork, which is the enemy of being productive.

Another useful mind hack is to give yourself a reason to work efficiently. It could be more time to hang out with your pals, or teaching your son to ride a bike, whatever works for you. The theory is the reason you give yourself will encourage you to work more efficiently, and therefore increase productivity. It will give you a new reason to work and will mean you don't work harder only so you can do more work. There is also a reward.

There is also a Ninja Mind Hack for people who have an issue with getting started. I used to suffer from this. I could quickly identify what I needed to do and why, but I would take plenty of time to get started without ever really knowing why. Perhaps it was because the whole project seemed overwhelming, and I knew it would take a long time to complete?

The method I used to overcome this was the Pomodoro Technique, which is similar to the idea of egg timers. The theory goes that you set yourself a time limit of 25 minutes to complete a task and then take a break. In this way, an enormous task will separate into small parts which seem much more comfortable to complete than a whole big project. Psychologically, you will find it much more possible to tackle a small project than the whole thing. After completing four Pomodoros, you will have achieved a lot, and it will be time to reward yourself with a more extended break.

It is far easier to get ready to do less than a half hour's work than for a full day's work. I found this an excellent method to overcome my issue with procrastinating. Just take the decision to do one Pomodoro, and at least you have made a start, then the rest will feel more manageable. Over time you will learn to predict how many Pomodoros a given job will take.

Some people are happier concentrating for a longer time and taking a more extended break. So experiment until you find the optimal work and break the pattern. The critical point is to use the Pomodoro method to break procrastination and make a start to your project.

A similar Ninja Mind Hack is the 30 - 30 method. It is when you do intense bursts of hard work for 30 minutes, and then you reward yourself with 30 minutes of doing something fun. It may be a viable alternative to Pomodoros for some people. The reward could be anything: surfing the internet, going to McDonald's, or going to the park - whatever is sufficient to motivate you. Once again, you can experiment with the time periods until you discover which is most suitable for you.

It would be best if you kept a journal of all of the things you accomplish per day. It will have the effect of giving you a sense of satisfaction and pride for what you have achieved. If you start your next work day by looking at this journal, you will feel motivated to achieve more. Subconsciously, you will wish to compete with yourself and do better than you did the previous day. It can be a powerful motivational tool for your mind.

Another thing to try is to have a change in your work location from time to time. It could be whatever you like. For me, I have always loved staying in hotels, even cheap ones. I love the feeling that I can order anything or get an issue sorted out with a quick phone call to staff who are a few meters away. So, from time to time, I like to check into a hotel and work from there for a few days. The change of environment does me some good, and I always get plenty of work done. When I go back to my natural environment, I feel refreshed and happy to be back, which, again, boosts my productivity.

Being unconventional can work as another Ninja Mind Hack. It can work as a way of making you feel as if you are different and can lead to you feeling motivated. It could be doing your household chores when everyone else is out partying or grocery shopping in the middle of the night. Whatever appeals to your sense of being unconventional.

Blocking Blue Light on your various devices is another useful Ninja Hack. Many of us have developed such an obsession with our devices that we can't do without them, even at night time. Studies have shown that extended contact with blue light from our devices can have a severe effect on our sleep cycle.

It is to be avoided at all costs (as we have discussed earlier), because a good night's sleep and feeling rested are one of the main factors in being at our most productive. Most up to date devices have an app to block blue light, but in case they don't you can download something practical. Whatever you do, don't let something so simple as exposure to blue light lead to unproductivity.

A straightforward Hack which you may not have thought of is to turn off all notifications of your devices. With no Twitter, Facebook, or email updates, you will be free to focus on the task at hand. You can assign a set period per day to deal with those notifications and use the rest of the time to get more things done.

A recent trend in Ninja Mind Hacks is intermittent fasting. It is an exciting idea which people believe makes our minds more focused and sharper after fasting for 12 hours. The theory is that at the point, our bodies naturally start using our stores of glucose, which has the effect of producing ketones, which are useful for encouraging brain metabolism. It could well be that new Ninja Hacks for our bodies could be the key to achieving higher productivity. It is an area which is worthy of further study and experimentation.

Consider using a "mind dump." It can be useful when you start to feel stale and frustrated with the job you are working on completing. Just write! Whatever comes into your mind, without worrying about grammar, spelling, or structure. Just let whatever is on your brain out. Let it go! You may feel refreshed after (even if what you wrote is nonsense!) and feel more able to get down to doing some proper work.

We all get stir crazy sitting in our offices sometimes. If you start to feel that you can't get anything done and are being unproductive, go outside. Break the pattern by doing something different for a while. Go for a walk, a hamburger, or even a beer. Just find a way to break the cycle. When you get back to the office, things will be different.

A short change of scenery and just letting the mind think about something else for a while is often the cure for when you feel that things are not getting done. Just be sure to tell someone where you are going, so that you don't experience the same thing I once did. Some emergency happened when I slipped out for an hour, and I was the only one able to deal with it. I returned to panic and chaos as nobody had any idea where I was!

The final Ninja Mind Hack I will recommend is Time Stretching and Attention Training. The basic concept is to have a period each day where you allow your body to go into a time of intense focus on something other than work. It would be meditation, competing in sport, playing an instrument, or working on an artistic project or some similar endeavor. The focus and effort you put into this practice will make you more perceptive and have a generally positive effect on your health.

All of these Ninja Mind Hacks will be beneficial in increasing your productivity and enable you to work smarter. You should add as many as possible to your day to day practices as is possible.

Secret Habits to Declutter and Get Peace of Mind Quickly

As we have discussed elsewhere, decluttering is a step-by-step process which can take a lot of time. However, peace of mind is helped by decluttering and it is an essential part of being productive.

In the following section, we will look at some quick methods we can use to achieve peace of mind in a matter of minutes. All of these are effective ways to improve your state of mind - quickly!

- It would be best if you went through the pile of old mail that you have been accumulating for months.
  - Quickly go through your house, collecting all old mail, flyers, and other useless papers. After checking there is nothing essential - junk the lot!

- Everyone has a drawer somewhere which is full of junk. It is time to tackle it! Take it out and turn it upside down.

- Only put back things which are useful and serve a purpose. Junk the rest!

- Do a laundry run and then neatly sort it out and put it away.
    - Many of us regularly feel the laundry building up and threatening to take over the house. It is time to get it done!
    - Once it is all clean and put away, a weight will lift from your shoulders. Make a promise to yourself, never to let it build up again.

- Look through your bathroom and throw out everything that you don't need.
    - The bathroom is another area where we collect unnecessary clutter. Old newspapers, magazines, shampoo bottles, and others are redundant. Why are you keeping them?
    - Another room has had a decluttering makeover in a matter of minutes.

- Look through your closet and throw out your old clothes.
    - Most of us have a closet full of clothes we haven't worn for years. It is time to get decisive and junk or send to the charity shop what we don't need as a part of the renewal process we are undertaking.

- Clean out your refrigerator and throw out everything that you won't use.
    - Firstly, remove everything inside and then give the refrigerator a good clean. Only put back things which you are sure you will use.

There is a load of clutter to dump using this method, and the best bit is it won't take long! Small steps matter in the overall process of becoming more organized. Doing quick decluttering in this way, and promising yourself not to let clutter build up again, is an essential step to becoming the most organized person in your group.

Every unnecessary item we own will work on our mind in some way. Taking these quick and straightforward steps is the right way of clearing your mind as well as your home.

Most of these things only take a matter of minutes, so why not do them today? Psychologically, you will feel better and ready to become more productive.

Organize Your Household Papers

We have all been in the situation where we are desperately searching through a pile of old papers looking for something important. What do we do when we can't find a bill due today?

It is time to get organized and sort out the household documents as part of the decluttering process and becoming more organized. You may be amazed at how many documents you find stuffed into drawers or closets, or just sitting on counters or tabletops.

Here is a proven method to get this issue sorted out, and once it is, we should make a pact with ourselves not to let it happen again. Promise yourself, write it down, and follow your promise!

- Round up all of your documents and put them all into one pile.
  - It will allow you to go through them and separate the important ones from the junk. Be brutal and throw out anything which is not essential. There is no room to hoard junk!
  - It would be best if you took advantage of finally doing this job to set up a proper filing system for anything important.

- Have a look through your documents to see what you can digitize.
  - It is time to get more high-tech with your simple household documents. Most smartphones or tablets can download a scanner app which will allow you to make digital copies which are easy to store.
  - Aside from any legal documents, everything else should have a digital copy made and then discarded. It means more clutter is permanently gone!
  - Make it a habit to scan, save, and discard any new non-legal related household documents you receive.

- For any essential documents, you need to have copies made.
  - For anything important like land titles, birth or marriage certificates, wills, and the like, it is essential that you make a backup copy and put the originals in a safety deposit box or similar at another location.

- If you were to lose these documents due to flood or fire, it would be time-consuming and expensive. For your safety and convenience, this is a critical step to take.

- Buy yourself a filing cabinet and set up a proper filing system.
  - You will be amazed by how much stress having a proper filing system will save you.
  - Buy a proper filing cabinet and other supplies such as files, labels, and others. You can then divide everything into categories such as bills, receipts, warranties, personal documents.
  - If you do not have too much in the way of relevant documentation, you may be able to get by with a large folder. As long as they are all filed away, you are making proper progress!

- Try to reduce the amount of paper you use in the future.
  - After you have scanned and saved as much as you can, you should investigate other ways to cut down on paper.
  - See if you can sign up for digital versions of household bills, bank statements, and others. These are all far easier to archive and store online than if you are collecting more unnecessary paper.
  - Over some time, you can put into place a system which will enable you to do a significant amount of decluttering and take steps toward getting organized.

- Taking control of your household documents and finally getting them sorted out is an integral part of becoming more organized, both mentally and practically.

It may seem like a small step, but it is a significant one. There is no reason not to start today. Being more organized is an essential step to becoming the most productive person in your group!

Make Sure You Keep Up your New Habits Which Have Allowed You to Become Organized.

All of these methods of improving your organization are not much good if you cannot maintain these new habits. If you fall back into your old ways every time, there seems little point in trying to become more organized.

I will share a few tips on how to retain your new good habits.

- It would be best if you started with something simple and not try to change too much at once.
    - Find a simple first change to make and stick to it. Turn it into a habit. An example here may be to get into the habit of digitizing all of your household documents.
    - If you try to make too many changes at one time, you will likely not stick with them long term and not form positive organizational habits.

- You should celebrate (in a small way) if you succeed.
    - If after say three months, you have been able to keep up your good habit of decluttering, you should reward yourself.
    - Becoming more organized is something to celebrate, even if the habit you have picked up is something small.

- Realize that you won't always get it right and mistakes are possible.
    - To err is human, so don't get too down on yourself if you break some of your excellent new habits.
    - If you slip up - don't give up! There is something worthy about having a decluttered house or a clean work area.
    - Whatever it is, please pick up and start again without letting things get as bad as they were before. Get organized once again and double down on your determination to follow good habits and be disciplined.

Adopting these steps I have suggested into your routine is an excellent way to develop positive new habits and a critical step toward becoming a more organized individual.

Don't Make These Mistakes When You Try to Become More Organized.

The goal of becoming more organized and trying to declutter your life to become more productive is a worthy one. However, there are a few critical steps to avoid.

- Don't invest in new storage before you start.
  - You may think that buying storage boxes, cabinets, files, and other items before you start is a good idea, but it is not.
  - Do the clearing up and decluttering first and then calculate what you will need for storage.
  - If you get the wrong storage things in advance, you will have a beautiful collection of new clutter, which defeats the whole point of what you are doing.

- Don't be too ambitious and take on too much.
  - Decluttering and organizing your home is a big job and will take time to achieve. Do it in an ordered manner so that you don't lose motivation.
  - A few hours of work which does not start to feel tortuous is better than doing a whole day and ending up hating it and giving up.
  - Small but smart steps are best. Just aim to do what is realistic. You likely can't do your whole house and office in one day. Take your time and plan for it to be a long term project.
  - You should feel good about what you have achieved, even if it is just a small step. It should keep you motivated to keep going until everything is complete.

- Make sure you complete the task.
  - You will likely have a lot to do to complete decluttering your whole house. Be realistic and set a time frame to achieve this.

- Decide what you will do with the things you will throw out. Surely, there will be much garbage, but separate the useful items that you no longer need. Some could go to charity shops or friends.
- One crucial step is that once your decluttering is complete, you need to complete the task by disposing of everything.
- There is no point separating everything into separate piles or bags and leaving it there. It defeats the purpose of what you are doing.
- At the end, take the time to throw the garbage, drop by the charity shop or deliver something useful to a friend.
- Finally, the clutter is gone, and disposing of it will have its mental rewards. It is an achievement!

- Don't fall into the trap of thinking this is a one-off job.
  - Decluttering and becoming more organized is a long-term task. You will need to repeat this unpleasant process many times, but hopefully not on such a large scale.
  - You need to be determined to be disciplined and not let the unnecessary clutter build up again. You are now an organized person!
  - Your organized new home will look great, and you will feel a sense of having accomplished something by decluttering, but make sure you are prepared to do it again.
  - If you want to be the most organized person in your group, this needs to be a permanent change.

- It can take over two months for a new habit to become second nature. So, expect to be doing plenty of more clearing up during that time.

- Don't expect perfection.
  - Things will inevitably build up again from time to time. We spend a large part of our lives in our homes, so it is natural that we may collect unnecessary junk.
  - Don't be disheartened when it happens. We are human, and after all, we aim to become the most organized person in our group - not to become perfect!

These secret habits and Ninja Life Hacks that I have shared with you are an excellent way for you to become a more efficient and productive person. You need to be disciplined and do it. If you can implement the majority of these into your day-to-day life, I have little doubt you will be by far the most organized person in your group of family and friends.

Chapter 7

How to Manage Your Daily Actions to Become More Productive and Achieve More

There are many ways in which we can more effectively manage our-day-to day actions, which will lead to is becoming more productive and achieving more. In the following section, I will discuss various ways for us to achieve that.

Firstly, I would like to discuss in greater depth the fantastic opportunities provided by technology to increase productivity. In my opinion, the invention of the internet and the accessibility to new and up-to-date information it has provided us has been very valuable. The web is the single most significant advancement we have made in history.

How to Easily Organize and Declutter Your Technology

Getting rid of e-junk is a constant battle for most of us. The popularity of cleaning apps for our phones and tablets is a testament to that. There are several favorite apps which perform this function.

Some of the more well-known include Phoneclean, CleanMyPhone, and Tenorshare for iPhone and Clean Master, DU Speed Booster and CCleaner for Android.

These are simple examples of ways in which we can declutter the devices which have become such an essential part of our everyday lives. The unnecessary accessories and data stored on our devices can be comparable to the inessential data that we store, which prevents us from working at our optimal level of efficiency and productivity.

There is certainly a popular trend for many people to declutter their lives and homes. Downsizing and throwing out unnecessary possessions is becoming popular. There is plenty of unseen data in our technology, which we should also consider purging. There is a vast amount of digital junk from old work reports, long-forgotten screenshots, and photographs which haven't been looked at in many years which we can afford to lose from our devices.

There is also plenty of hardware cluttering up our houses. We must all have multiple old phones and laptops hanging around unused. Our homes have wires, headsets, and cables which have lain unused and forgotten for ages. Why do we hold onto such pointless junk? The instinct is probably to think that it "might come in handy for something one day."

I am a gadget freak. I spend practically all of my spare money on yet another smartphone, tablet, or some accessories. I think I am uniquely qualified to comment on collecting junk. The question is, what we should do about it?

In my opinion, the biggest culprit is the various power cables we accumulate for each different gadget. For example, all of my laptops smartphones, cameras each require a different type of wire, which is annoying. My penchant for collecting the latest interesting gadget means my collection of cables grows by the month. The best way to handle this is to have a tech decluttering session. Gather your whole collection of cables and be decisive.

Brutal action is required, and you should throw out everything except what you use. Nobody is ever going to need the Nokia 3310 charger which has been in a drawer for ten years! If you don't know what it is used for, it is safe to say that you can throw it out without too many worries. Accessories such as these are so cheap to order online these days that anything you have accidentally thrown out you can easily replace.

Similarly, with old cell phones or tablets, if you have not used them for an extended period of time, it might be time to donate them to a charity shop. In our attempt to become technologically organized, we need to stop the disorganized spread out over the whole house. It is time to choose a drawer for everything tech-related to living!

Digital hoarding should be our next target. The data that we leave behind on our gadgets takes up valuable space and makes searching a more complicated process. We need to be disciplined and every once in a while spend a little time losing the junk on our laptops, phones, and tablets. It is not complicated and is quick to do. I recommend doing this at the end of each month, which will make the battle against e-junk more bearable.

Photos are a massive user of our limited data, purely because most people take so many. Many people never delete anything and think of it all as a fond memory. Perhaps we should be more particular in what we keep. Is the pic of the great hamburger we had five years ago necessary to keep?

This decluttering process is just as crucial to our mental process as it is to regain some space on our phones. We are carrying around our clutter with is on our phones. Addressing all of this is a normal part of managing our daily lives. If we can solve the clutter in our heads in a similar manner, we will make progress towards becoming highly productive.

Make Good Daily Use of Up to Date Apps to Become More Efficient

There are so many useful apps being rolled out on a seemingly daily basis that it is time to incorporate the best of them into our everyday business practices to aid with productivity. It is time to stop just using apps for fun and contacting friends and choose helpful apps to help our work rather than as a distraction. The time is now to use our devices for good!

There are just too many to review in an overall manner so I would like to take a look at a few apps which I believe to be particularly useful. It is worth spending some time going through your chosen Android or iOS store to check what is available and applicable to your business regularly.

Some of the apps I recommend you investigate further are:

- 1Password
  - 1Password is a useful app which can save us time by storing all of our passwords in one place. It is 100 percent secure and will assist you in selecting strong, hard to hack passwords.

- Grammarly Premium
  - Grammarly Premium is an excellent app if you work in a business where it is vital to get grammar, structure, and spelling 100% correct.
  - It will save you a tremendous amount of time, as it acts as a lightning-fast proofreader and will make suggestions about how to make improvements. For a relatively low investment, your work life will be more comfortable.
  - It is like a super powerful version of autocorrect. There is also a free version which is useful for testing the app or if you only need it for very light use.

- Google Drive
  - Google Drive is an excellent app, especially for those that need instant access to files, especially when working on the road.
  - Instant access to all of your company's files and documents from anywhere with an internet connection is a dream that would have been unimaginable for most people ten years ago.
  - There are free options and a very reasonably priced version for hefty business users.

- Google Docs
  - Google Docs is an excellent free program which is ideal for writing any documentation. You can log on using any Android or Windows device through your Gmail account.
  - I use this app all of the time, and I find the smooth interaction with other Google apps such as Gmail

and Google saves me much time and helps with my productivity.

- Hours Time Tracking
  - ○ Hours Time Tracking is a useful iPhone app which helps with keeping track of the time, scheduling, and using a virtual calendar.
  - ○ An added advantage is that it is a free app which integrates well with an Apple watch.

- OmniFocus
  - ○ OmniFocus is another Apple app, which is like a genius version of a to-do list. It is available in free and reasonably-priced versions depending on your needs.
  - ○ The big difference here is that it is a smart app that can set reminders depending on your location or with whom you are interacting. It certainly is a way to improve efficiency and boost productivity.

- Paper
  - ○ Paper is an Apple app which allows you to convey your ideas to work colleagues efficiently. You can sketch drawings, write diagrams, or make handwritten notes to show or send to others.
  - ○ It is excellent for sketches as it utilizes a touch screen and is so easy to draw or create on. It is a unique way to speed up the creative process and quickly share your ideas with others.
  - ○

- Click up
  - This app is available on both Google Play and Apple. It is an effective way to combine your to-do list with the agenda of your team.
  - It enables you to see the status of projects, the next projects which need to be done, and comment on the state of other colleagues' work.
  - You can easily share ideas, notes, or files and it is one of the better productivity apps that I have used. I recommend you install this free app now!

- Discord
  - Discord is a cheaper, but equally enjoyable alternative to the popular Slack app. It is used as a communication tool and is particularly useful for collaboration on a project with remote colleagues.
  - Team chat is a simple process as there is voice communication for multiple users, and the most expensive option only runs at $4.99/month.
  - If you require an effective and economical communication tool, you should seriously consider Discord.

- Be Focused Pro - Focus Timer
  - Earlier on, we have discussed the need to be focused and clear our minds to concentrate on a task to be truly productive. This Apple app is an effective way to help us to achieve that.
  - This app blocks distracting apps while you work on a task and allows you to set specific tasks which need to finish within a particular time.

- One great feature is that you can specify break times so you can set productive work intervals. You can also set work targets for a whole day.
- It is a competent app to help with work targets, especially for those of us who work well with work then a break. It is also customizable to schedule meetings into your workday.

- 24me
  - This app works as a personal assistant to sync all of your calendars together. These could include, Yahoo, Microsoft 365, Google, and whatever else you may use.
  - It gives you greater control over all of your work and personal tasks and makes scheduling a far more straightforward process. It is available for free on both the Apple and Android platforms.

- Dropbox
  - Dropbox is a popular app which is available on both Apple and Android platforms, which allows for secure document storage and sharing.
  - It allows you to organize and comment on files in a straightforward way. It is excellent for mobile devices, and you can check complex documents on your phone while you travel.
  - It is also easy for your team to stay in touch and make comments on each other's work. It is an excellent tool to increase teamwork and productivity.

- Evernote
  - Evernote is an Android and Apple app, which has the tagline "Meet your second brain."
  - It does indeed feel like that when you start using it. Most of us have millions of creative ideas running through our minds, which all too often get lost and forgotten.
  - It is a great way to save ideas on whichever device you have to hand, however random they may be.
  - This app is available in free, basic, and premium versions and is worth checking out for the most creative thinkers amongst us.

- FocusList
  - We discussed the Pomodoro technique at length in a different section, and this app is ideal for helping you follow that.
  - The basic idea is to use the technique for time management by using 25 minutes of intensive work, followed by a 5 minute break to be at our most productive.
  - This app helps us to plan this time management, and it also allows us to review how effectively we've worked at the end of the day.
  - It is one of the most effective and easy to use Pomodoro timers available on the market.

- Forest
  - Forest is another free app which is available on both Android and Apple platforms. It is a time

management app which focuses on blocking distractions.

- o It has a fun concept which turns discouraging you from picking up your phone into a game.
- o You are given a seed to grow a tree and you are rewarded with your tree growing if you leave your phone alone while you are supposed to be working.
- o Your tree will wither and die if you pick up your phone too often.
- o You can compete with your family and friends. Surely nobody will want to be the bad guy who lets a tree die?
- o The concept is a good one, which, in a fun way, will encourage you to leave your phone alone while you should be working and, thus, help you be more productive and achieve more.

- Lumen Trails
  - o This app is a free Apple productivity tracker. It allows you to track things such as calories consumed, what you do, what you spend, and how much sleep you are getting.
  - o Usually, you would use a separate app for all of those, but this brings all of those functions together into one app.
  - o It gives you feedback on your habits through quick notes and helps you identify problematic issues and encourages you to focus on increasing productivity.

- Pocket
    - Pocket is a dual free Apple and Android app which helps you to bookmark and store new articles to come back to later.
    - It is particularly useful if you are researching a project and are finding interesting information which you do not have time to read thoroughly at that moment.
    - You can come back to it later, and it will be presented on an easy-to-read interface. You can also grade how essential or interesting each article is so that the app will rank items for you.
    - They also offer you a reading list which will be suitable for your interests.

- Productive Habit Tracker
    - Productive Habit Tracker is a free Apple app which encourages us to follow habits which we hope to do every day.
    - You can add habits you wish to follow, such as writing in your journal, taking Fido for a walk, going to the gym, meditating, or writing a daily report.
    - Whatever you wish to get done, the Productive Habit Tracker will send you reminders and encourage you to get them done.

Whether you have a preference for Apple or Android, there is a vast range of useful apps available which will help you improve your efficiency and productivity. It is a current reality that you need to take advantage of these and boost your efficiency.

Your competition is most likely using something similar, so don't allow your opposition to gain an advantage in productivity.

Vow to Take Better Advantage of e-Commerce

I have been hugely impressed with the fantastic developments with e-commerce. There seems to be no limit to what is now available online. If, for a moment, we consider online shopping, we can now order practically anything we can find in the shops, often at a much lower price. Delivery is quick, and modern technology means there will not be any mistakes. When we consider how time-consuming traditional shopping is, it seems productive to shop via e-commerce.

The old doubts about using credit cards online, and fear of using your real identity for online transactions are mainly in the past, so there is an excellent opportunity for us to save time by shopping online. We would be foolish not to use the time saved to do something more productive. After all, there is always plenty of more work to do.

E-Commerce, more generally, has opened up so many opportunities in other areas. There is often no longer a need to send bills and send someone to collect the money. Why? It can all online. People can establish whole businesses, selling a wide range of stock without ever renting a building, or holding even 1 dollar worth of stock. There are options for the automatic answering of client inquiries, collection of leads, and automated billing, which will completely change the way we do business.

The future for e-commerce seems unlimited, and anyone who wishes to be more productive would be foolish not to stay fully informed of the innovations and learn to take advantage of them. It is a remarkably useful tool for increasing productivity and efficiency; and the influence of e-commerce will only spread in the future. Making e-commerce a day-to-day part of running our businesses will undoubtedly lead to us operating more efficiently and productively.

As we have looked at, there are various options to help us manage our daily operations and become more productive. We need to find the best combination of these methods for the business we are involved with, and they will surely help us to work more efficiently and achieve more.

Chapter 8

How to Triple Your Productivity Overnight With One Simple Strategy

It may seem impossible to triple your productivity in a short time, but the fact is that it is possible. Being more productive is mostly a state of mind, and there are many ways to achieve his state. There is one particular strategy which I have found most effective, and it has allowed many people I have taught it to to be able to increase productivity dramatically overnight and me.

It is mindfulness. The idea is that you are consciously aware of what you need to do and actively consider what you can do about it. If you concentrate on improving your state of mind and your intention to be productive, things will improve. It would be best if you decided to be more productive.

Mindfulness has become increasingly popular in recent times. Indeed it can now be described as being fashionable. It can be described focusing all of your awareness on what is happening at this moment. It enables us to focus on the present, the "now," if you will, and be aware of precisely what we are trying to achieve.

This may seem like such a simple thing, but when we are mindful and focused on our thoughts, we realize that we spend far too much time considering the past and the future. We are all too prone to daydreaming about trivial matters rather than focusing on the task at hand. How we deal with this is key to becoming more productive.

We need to find a balance. Many of us naturally spend too much time having our thoughts focused on the future, the past, or the present. It is not productive to spend too much time in the past or the future, or to focus so strictly to the present that we fail to learn the lessons of history, so we need to seek balance.

When we use mindfulness effectively, it allows us to be thoughtful of what is on our mind and the present time. It is a habit that we are well-advised to adopt as we have a likelihood of focusing too much on what we will be doing next week, next month, or next year. Using mindfulness will stop our brains' wandering and allow us to focus on the now and become more focused and productive.

What are the Benefits of Mindfulness

Having better focus is one of the main benefits of being mindful. Our minds being distracted is one of the most significant challenges to becoming more productive. While we are trying to focus on the job at hand, our brains tend to lead us to think about the hundreds of other things we need to do. We may feel obliged to check our email or walk out and talk to the sales team, just due to a nagging feeling that there is something to be done there. We may then jump to thinking about needing to take our child to soccer practice later and that our partner expects a nice dinner tonight.

Frankly speaking, our brains are always looking for a distraction and are prone to wander. Mindfulness is one way to improve focus and increase productivity. It is the best way we have to bring ours mind to order and get focused on the now and allow us to finish the task at hand. Mindfulness also allows us to practice better planning. These two things complement each other perfectly. If we plan, we can focus less on what we are concerned about and this will allow us to practice mindfulness more effectively.

A proper plan and a schedule for everyday duties are straightforward if we apply mindfulness. It can enable us to find an appropriate place for what we need to do day-to-day and leave time for focusing.

Reducing stress is a significant thing that mindfulness can help us achieve. Too much stress is caused by thinking about the possible negative consequences of our actions. If we can focus on the job at hand, we will not suffer stress due to this. Unfortunately, our brains are wired to speculate on the future, so it is hard to overcome stressing about this. Therefore, we need to find methods to manage this issue. There is no benefit to stressing over possible future issues which haven't happened yet. It is a ridiculous habit.

Mindfulness can help us realize this and bring our stressful negative thoughts of the future back to the present and understand the future should not be having an effect on the now. This use of mindfulness can be beneficial in dealing with insomnia caused by worrying about the future.

In my younger days, I suffered greatly from insomnia. It turned me into a physical wreck. Sleep deprivation is not helpful if you write for a living. I had been through years of misery with insomnia when a friend helped me realize that it was fear of what may potentially happen in the future, which was causing it. Mindfulness was what helped me overcome it. It made me realize there is nothing to fear.

We need to replace the negative thoughts about bills that we will pay and bosses that need to be made happy with more comforting ideas. Mindfulness can help us consider that things are not so bad. The bed is warm and our family is okay, our boss appreciates us and nothing too awful is about to happen.

You can use this moment to reflect that the future is not yet here, the past is past, and overall, the present is pretty good. Learn to appreciate and live in the now.

Fear and Mindfulness

Fear will change with mindfulness. You may believe that fear in various forms is stopping you from being more productive and effective. Anxiety can be the enemy. If we could choose between being fearful and fearless in our personal lives, most of us would choose the latter.

Many of us suffer from a lack of clarity about our decisions and reluctance to make risky choices due to fear. We should not have to feel restricted by this. We should not feel regret for being withheld from making bold choices due to an irrational fear of what might happen. Fear is a frightening feeling which often stops us from reaching our full potential. No matter how much we may try to avoid it, fear is something we all feel to some degree.

I used to suffer from many fears throughout my life, mainly when it came to my career. I worried I was not a good enough writer, I worried I might not make it, and I feared I might be broke. I stopped thinking about this and decided there was only one way to find out - work and see. I realized I needed to conquer my fear, and looked at various ways to find a way to overcome it.

Mindfulness is the solution. Realizing that fear will be overcome, the future is not yet here, and it probably will not be anything to fear at all. We also need to understand why we fear the unknown. A step into darkness always seems scary due to us not knowing what to expect. If what is in front of us is unknown, it will always be challenging to take the next step forward.

It is a very human reaction to be wary of what will happen next. We are predisposed to prepare ourselves mentally and physically for what may be around the corner. Emotions are a complicated thing, and they play a huge role in overcoming your fear of the future. Be mindful and base your feelings on reality, not an irrational fear.

Another consideration is how to increase your level of confidence. You need to understand yourself and your limitations before you can tackle your fears. It would be best if you aimed to be the best possible version of yourself, and an excellent way to achieve this is to boost your self-esteem and become more confident. It will help you overcome negative fears.

If you suffer from low self-esteem, the challenges presented to you in everyday life can seem insurmountable. Anything from making a speech in class to dating can become a traumatic experience. Low self-esteem can lead to a negative view of the world and even lead you to develop a "victim mentality." The world can seem a difficult place to conquer if you reach this state. It can lead to you feeling worse and your self-esteem falling further; and your productivity will suffer.

A boost in self-esteem is a crucial way to boost your productivity. If you suffer from low self-esteem, you will never be at your most productive. Fortunately, whatever level your self-esteem is at, there are various ways to boost your self-confidence. You will feel better overall, and your productivity levels will improve. In many cases, low self-esteem starts with how you perceive yourself and how you think you should be.

It can start in childhood with a thoughtless comment from a relative. The offhand comment saying you did something stupid or that you are fat can have long term consequences. The way that others treat or see you has a considerable effect on your self-esteem. However, the solution to this comes does not come from external factors; it must come from within yourself.

Improving your self-esteem is not easy to achieve; however, with the right methods and support, it is possible. There is a hidden power which can lead to a fantastic boost in productivity, if we can resolve our self-esteem issues. Some strategies to try include:

- Find out what is the real problem you have. There can be many triggers, and the use of therapists or simple mindful thinking can lead you to a greater understanding.

- Always do your best. It is a simple, but essential step. If you feel you are doing your best every day, you will feel pride in yourself.
  - If you are disappointed by an outcome to a project, ask yourself if you did your best. If you did, there is little else you could have done. You should feel satisfied.

- Try to see what others see in you. Positively do this by imagining how the one who loves you most sees you.
  - In my case, it would be my grandmother. If I see myself through her eyes, it is easier to see myself in a positive light.

- Do things that you enjoy. They are essential to give you a feeling of well-being and satisfaction.
  - When you enjoy what you are doing, whether it is hanging out with friends or a hobby, it makes everything seem worthwhile, including yourself!

- Accept the good and evil within yourself. Whatever has happened to you throughout your life, appreciate yourself.
  - Bad experiences always teach us things and make us stronger. Appreciate that you're well-intentioned and always do your best.

- Be aware of who you are and be proud of that. Identifying your real self and accepting it as a fact will lead to you improving your self-esteem.
  - You don't always need to "fit in" with the in-crowd. Learn to appreciate your good points, and accept your weaker ones, more.

- Compromise less and be stronger. Everyone wants to appear friendly but putting everyone else's needs before your own is not the way.
  - It will have negative consequences for you. It is necessary to realize that your needs are essential also.

- Always aim to look for the good in yourself. If your self-esteem is low, you will likely only see the negatives when you look at yourself. It has to change.
  - You have probably gotten into the habit of seeing your worst side. Be mindful of your good points and concentrate your thoughts on those. Your self-esteem will improve.

- Stop thinking of yourself negatively. Don't worry about the mistakes you have made. They are in the past.
  - If you tell yourself you always screw up and make mistakes, you will believe it. It's far better to say to yourself that you are capable and incredible as you will think it also.

- Value and appreciate your relationships. As our self-esteem is reliant on how others see us, make sure you surround yourself with positive and nurturing people.

- Appreciate people who know and love the real you. They will enforce a positive self-view by recognizing your better attributes.

- Take risks. A big part of being successful is experiencing failure. It will help you build resilience, character, and self-esteem.
  - When you overcome failure and achieve something, you are getting stronger. At the end of your days, don't regret not taking enough risks.

- Always have a goal in life. No matter how modest it may be, it is essential to always have an aim as each time you achieve it, your self-esteem will improve.
  - It is too easy to allow the fear of the unknown to stop you from achieving what you should in life.

If you follow these steps, they will go a long way toward improving your self-esteem and allow you to increase your productivity levels. One of the best things we can do to overcome fear is to understand our purpose in life. It will help us understand the unknown and conquer our fears.

Our sense of purpose can boost our productivity as it will increase the way we feel. A spiritual and emotional feeling of well-being can stimulate the mind to achieve more. Sometimes finding the meaning of our lives means more than happiness. Happiness is an emotion which can come and go; being mindful of our purpose can be a more effective way to be productive. The meaning to our lives will encourage us to be at our best during the good times and keep struggling through during the tough times. The meaning and purpose of life can help overcome fears as it allows us to know what our aims are and where we intend to end up.

Visualizing the future is your friend. It is a compelling method to rid yourself of fear. It is a way of rehearsing the future, to be better prepared for what you may face. Studies have revealed that visualizing what may happen has a similar effect on the brain as real action, so in a way, you are training your mind to be healthy and unafraid.

This practice can help you stay motivated, improve self-esteem, and put you on the way to achieving maximum productivity. It is another example of the power of mindfulness and allows us another way to improve our effectiveness. If you consider all of these points and act on them to use mindfulness, you can triple your productivity overnight.

The improved focus, overcoming your fears, better planning, and determining to be less affected by stress and its side effects are all critical components to becoming a more productive and happier individual. It is possible to become more productive in a short space of time. You should start practicing mindfulness today!

Chapter 9

The 3 Scientifically Proven Things You Need to Stop Doing Right Now in Order to Get More Done

We can all agree that we typically need to find a way to perform better and get more done. The trick is to find the most effective way. We will next consider the three scientifically proven things we need to stop doing right now in order to get more done.

The things we stop doing can be more important to our productivity than the things we are doing. We spend far too much time working on negative things and finding ways to eliminate them is vital to improving overall productivity. The medical community has been proving that attitude can have an effect on our health for the past 50-plus years. There are countless studies showing how harmful negative thinking can be.

The three scientifically proven things we need to stop doing right now in order to get more done are:

- We need to get rid of negativity in our lives.

- We need to stop constantly looking at our social media sites.

- We should pay more attention to our ultradian rhythms.

We will now consider these three scientifically proven points in turn in our effort to understand how we can be at our most productive.

How to get rid of Negativity in our Lives

Negativity is the enemy of productivity. We need to get out of the habit of saying "I can't," and become more positive and optimistic. The fact is, you "can" do anything, and if you don't know how yet, you can learn!

If you have a bad habit you wish to break, such as spending hours reading online when you should be working or an addiction to smoking, don't say "I can't" but find a way to make sure you "can."

A negative mindset will ensure that you fail. It is a trap that too many of us fall for. The way you think means the difference between success and failure. If you find yourself doubting your ability to do something, keep telling yourself "I can do this, and I will do this". If you truly want to improve your productivity, changing the way you approach every day challenges is an important way forward.

Having some negativity is a human way to be and is easy to understand, but having too much will just make you underachieve and make life harder. A pessimistic attitude will make everything seem negative and make it impossible to achieve the success we want. Make a vow to adopt a more positive attitude and enjoy your new positivity. You need to be proud of your strengths and view any weaknesses as just something you are working on improving.

Boosting your productivity will mean stopping negative habits. This may seem hard to do, but start now and you will see that after just a few weeks, your productivity will be improved. There are various ways you can try to reduce negativity in your lives and these are proven methods to try:

- Body language is important. It has more meaning to our mindset than you may realize. If you are slouching or frowning, you are more likely to be thinking negatively.

  If your body language is poor it can affect how people see you and this can have an effect on your self-confidence which in turn leads to negative thinking. Focus on improving this. Sit up straight, smile and be more confident when you deal with other people.

  You will start to feel better if you improve your body language, and this will help you lose your negativity; improved productivity will follow.

- Change the way you think. Sometimes changing your perspective of the things around you can help. Instead of thinking of everything as a, instead think of it as a challenge.

  There is a huge difference between thinking, "I will never finish this sales report by tonight," and "This is a challenge and it will be a real achievement to get done on time". The two are very similar, but the positivity in the second statement will allow you to banish negative thoughts and make a huge difference to your productivity.

- Some things need to be discussed to solve issues and get pent up emotions out. Keeping things to yourself can build up your frustration and lead to negative thought patterns. If there is something you have been holding onto which should be discussed, you should take the opportunity to discuss it with someone.

  A simple discussion and hearing another person's opinion can help you put things into perspective. This can help you realize that the issue at hand is not so bad, and you can solve it. This will help you overcome a negative attitude toward it.

- Take a short time-out to calm your mind. We all experience a cluttered mind from time to time and confusion leading to a racing mind. It can be hard to maintain control over your thoughts and stop negativity creeping in.

  Just take a minute to calm things down. Some people like to meditate, I prefer to have some candy and stare out of the window. Empty your thoughts and restart your task; you will find much of the negative thoughts have gone.

- Go for a walk. If you are stuck with a negative mindset, you're better doing something to change it. As negative thoughts occur in the mind, it is easy to conclude that is where they come from, but there are other factors.

  We can't always choose who we work with, and negative people can cause you to follow a similar thought process. Sometimes the best choice is to take a break from this atmosphere. Head to the park, people watch, or have a drink; just do something to change the negative vibe.

This will improve your state of mind and you will be more productive when you return to work. The short time off will be more than made up by your improved mindset.

- Do something creative. Embracing your creative side is an excellent idea to overcome negativity. Find a creative outlet, whether it is writing, painting, or building something. It will lessen your negative feeling. Creating is always a positive step.

This can act as a form of therapy as you can release your emotions through a creative process. It will help you reduce negativity and become more productive.

- Take time to notice the good things in life. It is too easy to forget the things we should be grateful for. It could be the nice lifestyle your job allows you or the love of your family, but most of us have plenty of things to be grateful for. When we get stressed by day-to-day life, we stop focusing on the things that are going well in our career and family life.

Just because your boss shouted at you or you had an argument with your partner doesn't mean it is all bad. Accentuate the positive. Physically making a list of all that you should be grateful for is a good start. It helps us focus our mind on the good things in life. This will stop us taking things for granted and failing to see the good things. We will forget negativity if we focus on the good. This is a powerful tactic to use in overcoming negativity and is certainly something you should try.

- Laugh at yourself. It is too easy to take things too seriously when life gets busy, so you should practice seeing the funny side. We all have our foibles and strange ways of doing things - we should laugh at them. If you don't take yourself too seriously, you won't get trapped into negativity.

  Learning to laugh at our mistakes and silly behavior will help us realize things are not so bad; and make us feel happy. Our happiness will lead us to positivity.

- We should strive to help others. Negativity and selfishness are related, and in order to find a purpose to our lives we need to consider others. If we can find our purpose, we will become more positive. It is the simplest way to become more productive.

  You can begin in a small way; be polite, hold a door open, or inquire how a colleague is doing. All of these will give you a sense of worth that will be helpful for positivity.

- Form your own team of positive people. There is no better way to overcome negativity than to spend time with a group of positive people.

In the same way that negative thinkers lead us to negativity, the positive people in our lives encourage us towards positivity. You should put your own support team of positive thinkers together. They can be relatives, friends, colleagues from work, or whatever; as long as they are positive people. You should aim to meet with them regularly, for a drink or a lunch, and feed off of their positive vibes.

This will have a great psychological effect on you and help you fight off feelings of negativity until you next meet up.

- Become more aware of your surroundings. Start to pay attention to the small things and appreciate the day-to-day things we deal with. Learn to appreciate the beauty of nature more, laugh at things more, spend time hanging out with your pet more. You should note how you feel doing all of this. This awareness of the interesting and beautiful that surround us is an excellent way to decrease negative thoughts.

- Learn to identify what makes you unsatisfied. Sometimes it is hard for us to identify what it is we want in life. An effective approach to this is to consider what we don't want. Most of us have something that we don't like in our lives. Perhaps it is a lack of skill or knowledge or a negative person who you know you need to lose.

Consider these points for a brief time and then focus on how different your life will be if you finally do something about it. You can use what you don't want to identify what you do. Positive changes will have a dramatic effect on your mindset. You will feel better and become more productive.

- Find out what motivates you. It could be increasing your prosperity or providing for your family; perhaps something as simple as an extra holiday per year. These thoughts will give you an aim in life and something to strive for. Test sense of achievement you will feel while building for your aims is extremely rewarding.

It will lead to positive thoughts and a "can do" attitude and when you achieve your aims, your sense of well-being will be boosted. It will help you to reduce your negativity and focus on what is possible if you try.

- Accept that you are still learning. It is easy to get trapped, in our career or our relationship, by a feeling that we can't improve our situation. If you can accept that we are not all-knowledgeable and are still learning, there is always hope that things will get better.

Take the chance to do some more studying, read a book, or even consult with a therapist. The learning process will help you understand your situation better. The extra knowledge which you acquire will help you know there is a better future and this will help you have a positive attitude. Knowledge is a powerful way to overcome negativity.

- Take a moment to realize how strong you really are capable of being. We all face challenges in life and it takes a great deal of internal strength to face up to them. We are all vulnerable to feeling sorry for ourselves, but when we consider our lives, we should recognize what an incredibly resilient machine we are every day.

Take pride that you can overcome difficulty so well. Focus on the fact that every hardship has made you stronger. This will allow you to improve your self-esteem and acknowledge your strong points. Human beings have an awesome power.

- Learn to enjoy your work day. Instead of viewing your job as a chore which has to be done, focus on the good things that it brings you. Even if you find your job boring and not much of a challenge, it does bring good things to your life. Think of what your salary gives you the freedom to do and the pleasure that you get from a chat with a friendly work colleague.

  One of the challenges I initially faced as a writer, was the lack of day-to-day interaction with work colleagues, so I consciously made friends in my field of work. Making the best of it and enjoying what you do is an important way to encourage positivity.

- You should take some each day to let your imagination wander. Turn off your appliances and consider your dreams and fantasies. It is a great mental process to let your mind run free, and I find I get some of my most creative ideas when doing this process. Your dreams and fantasies are something unique to you and will help in your quest to have a positive mindset.

- It is a good idea to consider how long your goals might take. We have discussed how having goals leads to a positive mindset. It is worth considering how long they might take. Short term goals help us feel good about ourselves when we achieve them. Something that will take many years to achieve might cause you to lose interest, so it is helpful to consider what is possible in the shorter term.

For example, you may have always wanted to learn a musical instrument. This is perfectly achievable in a relatively short time. The pleasure this will give you and the sense of pride you will feel when you achieve this will have a great effect on your mindset.

- You should always reevaluate your goals and find new ones. It is important to keep moving forward with what you wish to achieve. Nobody likes the feeling of stagnation; it leads to boredom and increases negative thought patterns.

You can set a time every month to consider how much progress you have made towards your goals and what you can do to improve your progress. There are always new goals to be set, and you should aim to add a few more every month. I try to set a goal to learn a new skill every month, even if it is something simple. As I am a writer, I aim to learn five new words every day.

These seemingly small steps help me overcome negativity, as having something to aim for focuses our mind on the positive.

- We should realize that our relationship habits have an effect on how positive or negative we feel. There are certain things such as being married, owning a pet, laughing, and having a loving relationship, which are scientifically proven to make us happier. These things can also have a drastically positive effect on our health.

  There has been plenty of research which shows the link between heart disease and not having strong links to family and friends. Loneliness can seriously impact our health and state of mind, so we should be conscious of this and consider what we can do to improve our personal relationships. If we have a positive relationship with family and friends, it will be difficult to get caught up in negative thought patterns.

  A nice side effect to this is many studies have shown that this can increase your lifespan. A long and healthy life can be attained with positive relationships and habits.

- You should learn to forgive and forget. One of the most beneficial things we can do to decrease negativity is to forgive. If we forgive others, we can get out of the pattern of negative thought. It is one of the most healing things we can do for ourselves.

  There is no point in holding onto old vendettas and worrying about arguments from long ago. You should just let it go. Life is too short to have hatred in your heart for something that doesn't matter much in the great scheme of things. Forgive and forget. Does the stupid bit of gossip someone said about you really matter that much?

Try this method. Forgive someone today. You will see that you will feel better. You will get rid of negativity and will become more productive.

It is scientifically proven that if you can stop being negative, your productivity will improve. Please take on these points I have raised to overcome negativity. If you implement what we have discussed, I am confident that you will be well on your way to creating a positive mindset that will make you a powerhouse of productivity.

Once your thoughts become more positive, you will feel happier and be better equipped to deal with the challenges that life sends you. It is beneficial if you can find the right balance in your life. Don't spend too much time working and give plenty of time to your family and friends. If you have strong relationships, your mindset will be stronger and positive things will start to happen.

When you become a happier and brighter person, life will seem easier and better opportunities and the opportunity for greater prosperity will come your way. It is time to stop struggling with negative thoughts and embrace the thought that human beings have a tremendous capacity to create a wonderful life. We all have tremendous opportunities in our lives and it is our duty to find a way to take advantage of them. Negativity will only work to stop us reaching our full potential.

We need to surround ourselves with as much positivity as possible, whether it is thoughts, upbeat people, good deeds or forgiveness. If we wish to be as productive as we can possibly be, we need to learn the lessons which we have discussed and act on them!

The new positive and productive you is well within reach. I advise you to start making changes in your life to achieve this today.

We Need to Stop Constantly Looking at Our Social Media Sites

The incredible rise in the influence the internet has had on our lives over the past few decades has had both positive and negative influences. Of course, the access to information and entertainment has led to us being better-informed and given us better access to quality entertainment.

I am a fan of boxing and I find it unbelievable that I can watch practically any fight worldwide live on my iPad. It was not that long ago that I would have to wait months to find out the results. So while there are clear benefits, we do need to consider the downside also. Addiction to social media is a real problem for many people. There is now such a wide range of choices of social media such as Facebook, Twitter, Instagram and Snapchat, that we are almost overwhelmed by choices to use.

There is a temptation to use all of them and this will have a devastating impact on our productivity at work. Even if we are not using it, we are thinking about it when we see something interesting. Even in our home lives, many of us find ourselves doing less, even spending less quality time with our families. There are millions of people worldwide who keep their social media accounts live 24/7. Is this really a healthy way to live?

It was not so long ago that we only used phones for talking to people. The long term effects of this are still unknown, but we have certainly undergone a drastic change in the last decade. While a lot of important things do happen, it is a fact that many of us waste a huge amount of time every day with frivolous things on social media. We should focus on limiting our access to social media, especially when we are trying to do some work. There is always too much temptation to randomly get chatting to a friend or watch another cat video.

It is a good idea to limit your social media use to one or two designated sessions per day, otherwise you will get distracted and your productivity will suffer.

We spend too much time interacting with people online, and we need to maintain real life relationships in order to properly bond with people and feel properly supported. While it is true to say that many businesses use social media in a positive way to promote their services or products, it does come with some risk.

The obvious risk is to do with time management. If an employee is tasked with using the company's Facebook account, the temptation to waste time by using their own accounts will always be present. It can take some time to refocus on the job at hand if you have started to look at some great pictures of your summer holiday location. It also disconnects the employee from interacting with others in the office and having a good work relationship.

There are many mental health professionals who fear that short attention spans and an inability to concentrate will be a side effect of prolonged social media use.

There is also concern about the effect social media use has on overall mental health. Many intensive users have reported that they suffer from high levels of stress. There needs to be a long term study as to the effect social media has on our mental health.

The mental health of your employees can impact directly how your business performs. You need content and motivated employees to be productive and have great relationships with others. A stressed employee can have various physical and mental issues which will affect their ability to have effective communication with clients and work to maximum capacity.

Social media can also be a source of misinformation. Fake news has become a popular expression and it is a fact that misleading online content can damage your business. Social media can also blow small things out of proportion. All companies or employees make errors from time to time, and if you have an incident go viral it can be a public relations nightmare.

There is also the Fear of Missing Out (FOMO) phenomenon, which entails a fear that you may be missing out on a fun or interesting experience that someone else mentions on social media. It is a form of anxiety which is fueled whenever you use social media and the more you use it the more likely you are to see someone else having fun you might not be having which adds to your stress.

Another side effect of spending too much time on social media is the negative effect it can have on how you sleep. I freely admit that I am a Twitter addict and I have to check it if I am awake for any reason. Even a trip to the bathroom in the middle of the night can mean going back online for an hour or more. It will eventually lead to sleep deprivation and this will certainly impact on our ability to work effectively and lower our productivity.

Cyberbullying is often thought of as an issue that young people face, but it is still a factor for adults even within a company. It comes in several forms such as nasty emails, fake photographs, gossip, and threatening private messages. Employees who have experienced cyberbullying often report extremely high stress levels which can affect their ability to perform in an effective manner.

There is also a potential for tension amongst colleagues who follow each other's private pages on social media. Jealousy can be a factor as can controversial opinions about politics or sports. Unfortunately, in the world we live in, opinions can divide us and cause problems. It can be a danger to harmonious relationships in the workplace, which can have an effect on team spirit and productivity.

So we can clearly see that there are negative side effects to using social media and limiting use of it is essential in the modern workplace. If you are experiencing any of the more severe symptoms, such as stress or depression, it might be time to consider giving up on social media altogether.

We Should Pay More Attention to Our Ultradian Rhythm

We all experience a drop in our productivity after focusing on a task for an extended period of an hour and a half to two hours. This is a scientifically proven natural occurrence which is known as the Ultradian Rhythm. It is something that we need to stop ignoring. When you feel mental fatigue, you need to ignore the temptation to keep on going and "push through" the lull in energy you are feeling.

You would be far better off taking a short break and having a cup of coffee or a snack, or better yet a quick sleep, before going back to work with a more focused attitude with more energy and creativity. Your body will use this time to refocus, recover, repair and rebalance. You will feel refreshed and capable of your next burst of activity.

A fifteen to twenty minute break can make the difference between an efficient day and a day of sloppy work. It can guard against stress and fatigue. What you definitely do not want to do is keep doing what you have spent the past few hours doing, which is probably staring at a screen.

You need to perform a "human reboot" by changing scenery for a while and concentrating on other things. Other activities you could try are; taking a bathroom break, going for a quick walk, looking at the sky, meditating or doing some deep breathing, yoga, letting your mind wander, or paying a visit to a friend.

In fact, any change of scenery and doing something different will be beneficial and make you feel better when you do get back to work. If we reach the state of feeling distracted and groggy, our immunity drops, our mental capacity is comprised, our metabolism goes haywire, and we will be irritable and moody.

In fact, we get less productive and the more of these breaks that we miss, and the more we risk damaging ourselves. The more you learn about Ultradian Rhythm, the better you will understand your own needs and be capable of improving your productivity and focus. This seems like such a simple thing to consider, but paying attention to your Ultradian Rhythm is one of the best things you can do to improve your productivity.

These three scientifically proven things which we need to stop doing are all effective in helping us get the boost in productivity that we are seeking. Experience has told me that each of these three steps will enable you to feel better about yourself and allow you to achieve the maximum efficiently that you can.

Chapter 10

Conclusions We Can Draw About Improving Productivity and Decluttering our Lives

I do hope you have enjoyed listening to my theories about how we can declutter our lives and seek better productivity in our work and personal lives. If you practice the techniques I have shown you, it will have a positive effect on your life and improve your situation at work and home.

A lot of the things we have discussed are common sense, and you are likely to use them to some degree already. There is no harm in thinking more deeply about ways we can implement these productivity methods to a more effective degree. I recommend you spend a part of each day reflecting on what we have discussed and decide which of these ways of improving your productivity are most suitable for you.

As we have discussed, there is a lot to consider when we seek to declutter and improve our productivity. We have looked at various methods to improve our productivity, which has covered effective processes to improve our attitude and state of mind. Decluttering in all parts of our lives is a positive step to improving overall productivity, but it is not at all easy to achieve.

It takes a long term determination to change our habits and thought processes. It seems the human brain is wired to go back to old negative patterns. To increase our productivity, decluttering is an important step to undertake. It should not daunt you as we have been through various methods to help you achieve this.

Decluttering is more than a physical process and can have tremendous mental benefits also. The discipline involved can help reduce negativity and put us in a positive frame of mind. We can all agree that we have room for improvement in our lives, and finding the best way to declutter is a significant step toward becoming a more productive person.

We know that modern life is increasingly about competition. Improving our productivity is the best way to gain an edge in our competitors and achieve the success we aspire to have. We have discussed the most effective methods to utilize the power of positive thinking and work to improve our mindset to be as effective as we can be.

There is no single way for any of us to reach maximum efficiency, and we now have the information to make incremental changes and work on different methods to have an overall effect. There are many ways to gain control over a given situation. Mastering this is a crucial element to being at our best. If we are stressed or worrying about minor issues, we are likely to overreact, be ineffective, and achieve less.

Finding a good work/life balance is vital to enjoying our lives and getting into a good frame of mind. Negativity is the enemy of productivity. As we have seen, there are ways to change the pattern of negative thinking. If you conquer your negative thought processes, you will have a happier and simpler life and be a long way towards being at your most productive.

To have a successful life, we don't just need to consider what we achieve work-wise; we need to ensure our family is happy and what we can do to help society as a whole. These methods to improve your productivity and declutter your life have come from a lifetime of experience, and I hope you get the desired effect when you put them into practice.

I wish you positivity, a decluttered life, improved productivity, stable home life, and above all else, prosperity and happiness.

Project Management For Beginners: A Powerful System For Managing Projects, Planning, Organizing & Scheduling Work & Life - With Proven Productivity, Leadership & Procrastination Hacks To Get More Done

Introduction

To begin with, planning an event, a process or even a simple step is something that is synonymous to our everyday lifestyle. As the saying goes, he who fails to plan will definitely plan to fail; thus, the need for having careful and strategic planning and execution of our daily activities no matter how little or insignificant they might turn out to be. Now, this brings us to the world of Project Management.

Additionally, the paradigm shift in the organizational and professional side of the world today has called for a whole new dimension as regards effective and efficient output. Lots of companies, organizations, firms, and so much more now focus their lens on the effective planning process in order to achieve more within the shortest time. And no matter how well we turn this, Project Management seems to be the only available option here. Thus, the need for Project Managers.

Be that as it may, if there is a high rise of people rushing into Project Management as a result of this paradigm shift, then we can say there are going to be a whole bunch of ignorant and inexperienced people that would definitely find themselves in the world of Project Management for the first time in their life; thus, the need for this book.

I'm very sure most of you are familiar with these two words – Project Management. While some must have undergone a professional course on it, others might have read a thing or two about it online. Some might not even have the faintest idea about the words altogether. Nevertheless, this book will keep you abreast with the world of Project Management and how you all can infuse it in your daily life.

Do you find it very hard in organizing, planning, and executing a project, a simple task or even a daily activity no matter how hard you try? Do you find it hard to relate and work hand in hand with others in order to come up with an amazing result? Do you find it almost impossible to stay consistent in your line of work or business, thereby, leading to fluctuations in your success rate and inconsistencies?

Well, not to worry, this book would enlighten you on these individual aspects, thereby making sure you turn out even better than the way you were before picking it up. It will broaden your horizon in the area of Project Management if ever you've had an idea on the concepts earlier. It would still go ahead in making you see the recent trends that have taken the forefront of Project Management.

With Project Management, all your days of unbalanced management and planning are over. This book would go a long way in correcting such outcomes and in the end, you will be filled with more than enough confidence while planning and executing. There is no better feeling than knowing exactly what you are doing and having control.

Far from planning and executing, there are other key areas you would never find in most Project Management books and contents out there. All these important but rare segments and areas are well explained in the chapters of this book. Additionally, there are practical and real-life questions and answers this book would throw at you. Instead of keeping everything theoretical, this book would take you into the real sense of the world. With practical examples and real-life situations, this book would ensure you fully understand the concept of Project Management.

Be that as it may, all beginners out there are liable to make mistakes in the course of Project Management. When these happen, then I will urge you to stay calm and collected no matter how the situation may be because the solutions to these mistakes might just be right in front of you. This is why we would outline the key mistakes that are quite familiar with beginners. Learn from it and the sky would definitely be your starting point as regards Project Management.

In the end, no one will be able to surpass your Project Management skill, not even those that studied it as a discipline. We are going to mold you from being a beginner to an expert in this field. But till then, why not take a deep breath, relax, and allow us to take you on this exceptional journey into the world of Project Management? Let's get to it, shall we?

Chapter One

Project Management at a Glance (Pre & Post Historical Background)

The rise of Project Management in today's world was a gradual process which saw it skyrocket rocket from the least of the things cherished in the world of professionalism to the top of them all. Ripple by ripple, what had seemed like an ordinary idea conceived many years ago had gained its strong footing in the world as a whole. People now introduce the art of Project Management into whatever they want to do.

For example, Project Management had now become a necessary condition for prospective job seekers before getting employed. Employers also didn't just stay relaxed, they also followed the bandwagon by focusing their attention on this necessary condition. They believe planning before executing is the most important trait one can hold in the professional level, that way, their mind would be at peace whenever they give an employee a task to complete.

Now, this is just a tip of the iceberg as regards what Project Management has to offer. Be that as it may, it is important to know that Project Management in real sense deals with logic, tact, resourcefulness, and creativity. These are attributes even the most boring Project Managers must possess in order to make things work in their favor.

Project Management deals with seeing the other components and parts of a project as an extension or even a combination that makes up the whole. In order words, the different parts that make up your project should be seen as related and important ingredients that would make your project effectively managed. Thus, each unit, section, and the cabinet must be given due attention.

Let's say your whole body is a project and your hands, eyes, nose, legs, chest, toes, and other parts are all different but important units that make up the project. Would you neglect the eyes and focus on other parts? The answer is no. And that is the same way you won't neglect the nose or legs and expect to have a perfect and healthy body. This same instance works perfectly for Project Management.

You can't expect your project to be effectively managed when you don't even give each necessary component the same proportion of attention and care. You really need to make sure these different units and components are adequately managed too. That is the only way you can get a good result in your Project Management. As a beginner, you might get carried away or overwhelmed by a particular unit in a project and often focus your attention more in it without balancing it with the rest. Now, that is not something bad, especially if you are able to detect this problem sooner. However, if you weren't, then don't beat yourself up about it. Trust me, you can always pick yourself up again and again.

Now, as a beginner, I know you must have been dying to know all about the historical background of this amazing way of getting an effective result and organizing oneself. Also, if you are an expert, it is quite pertinent for you to stay abreast with the historical background of your cherished Project Management. In case you start saying you don't need to know the history or you aren't interested, then, believe me, knowing the history of Project Management would come in handy when the need arises. Additionally, knowing the past would definitely open up our mind for better digestion of the future. In order words, if we know the history of Project Management, we would be able to fully understand and tweak Project Management to our favor.

Historical Background of Project Management

Categorically, we wouldn't be far from the truth if we boldly say that Project Management has been around for like an eternity. Come to think of it, planning, executing, and management is three things that have been quite synonymous to man right from the inception. Even the caveman or the early man would make traps in catching their prey and food. They would create shelter for themselves. Create a fire in time of cold. Now that is what I call planning and management.

What makes theirs different is that the name 'Project Management' hadn't been conceived and instead, 'Survival' was used in place of it. Be that as it may, it's still called Project Management. You can't seriously say because the name wasn't introduced then, thus, it's far from Project Management. Hope you can share the same thought for Apples, Grapes, Blueberries, and so much more.

It is important to know that there had been lots of important, massive, and ancient structures in time which was only possible as a result of proper and adequate management. A very big example is the Egyptian Pyramids. There is no way anyone could pull that off without assembling a group and managing them adequately. Now, that is Project Management at its peak.

Thus, what picture am I trying to draw here? Project Management is real and it had been in existence since the beginning of time. It is not something new as many would have you believe. It is, in fact, one of the oldest ways of making things done effectively and efficiently without wasting resources. Following this amazing work of collective effort, many philosophers and scholars now began attributing and associating the feat with the term, Project Management.

Ripple by ripple, Project Management became codified and also became a discipline which many would come to study. Along the line, many vital points and ideas were propounded as a contribution to the development of this new discipline, Project Management. A very big example is the Gantt Chart and the Agile Manifesto. Ever since its development, it had continued to make a reasonable impact on the lives of many. Though it might be an old process of getting things done, it is still as effective and efficient as ever.

It is important to know that modern day Project Management is very different from the ancient one. While the ancient system of managing projects is said to be crude and rusty but effective, the modern day Project Management is well detailed, well structured, and well organized. Thus, making their foundation and tenets a bit different though still the same in some ways. For example, when the Egyptians built the large pyramids for the Pharaohs, they knew such project would be enormous and takes time, thus, they assembled a large group of people for the task, but not without having someone among them as their manager.

According to the records, each manager and their group of hardworking people were entitled to a pyramid and must be completed within the same time range. These managers showed great enthusiasm, character, and commitment towards the project and ensured they did a great job in the end. Now, those are part of the key attribute you must exude as a great Project Manager.

Furthermore, when the great wall of China was erected, the emperor knew it was no small feat. Thus, there were records that show the intense planning and executing even before the project began. In today's world, one should be sure to plan his or herself beforehand, so as not to be taken unaware or caught off guard as regards any project. If the Old Chinese Empire hadn't planned ahead of time, they wouldn't have been able to complete or even start building the great wall.

Additionally, a large group of people was tasked to complete the project. Ranging from the commoners to the elites, from the criminals to the soldiers, from the young to the old, from the rich to the poor. In total, millions of people were tasked to start and finish up the project and not without station managers placed at every strategic point of the wall. Well there you have it, these are evidence that shows how much of importance Project Management had been over time and how long it had been of existence.

In recent times, everything we do points towards Project Management. Every one of our key industries today needs effective Project Management in order to get things done. From the Manufacturing Industry down to the Construction Industry. Everything boils down to Project Management. When Sir Henry Ford developed his world record assemble system which made his work 10 times faster, he was basically applying the knowledge of Project Management.

When the American states embarked on rehabilitation and reconstruction after the Civil War, Project Management was what helped them in making sure everything was done effectively and efficiently while mincing every resource out there. Project Management goes way beyond discipline, it is a part of us. It is what we do every day of our lives either consciously or unconsciously.

This is why as beginners of this amazing discipline, we should always look beyond the ordinary. That way we would realize that with Project Management, everything around us would definitely take good shape. Notwithstanding, we would definitely find ourselves at the top of every situation; either good or bad. If we look at every situation closely, be it a task, a project, or just normal activity, we would be stunned at the fact that it's either there is a leadership role to be played, a particular budget put in place or a schedule to be met at all cost. However, with Project Management, successful delivery is definitely guaranteed.

Modern Day History

It is important to know that Project Management was coined into its modern form with lots of charts, theories, formulations, and principles introduced at the 1990s. With time, technology started taking a whole new dimension with lots of exciting innovations and inventions being developed by diverse companies all over the world, so also Project Management. In order words, there was a need for the management of these amazing project. Here are some of the few key contributions and ideas of modern-day Project Management;

1.   The Principles of Scientific Management: Frederic Taylor was the developer of this principle in his 1911 publication, "The Principles of Scientific Management". In his work, he focused his lens in the steel industry where he hoped to transform the unskilled workers into more complex form by striving hard to learn new and simple techniques. He also mentioned the importance of having incentive-based wage systems out there as well as the general use of time-saving techniques.

2.   The Gantt Chart: As a beginner, you would definitely come across this chart as many times as possible. Many believe that the Gantt Chart is the foundation of modern-day Project Management. Little wonder why they also see Henry Gantt as the father of modern-day Project Management. Be that as it may, the Gantt Chart is an innovative and eponymous diagram which can be said be as effective and efficient as Project Management itself.

According to Gantt himself, he believes that the Gantt Chat would allow you visualize tasks ahead of time and even enable you to link these tasks together. The importance of the Gantt Chart is for one to be able to keep their schedule intact. The Gantt Chart had been used in executing top projects all over the world since its introduction. For example, the Hoover Dam built in 1931. In recent times, it has even shifted it's focus and attention in the digital world as it now comes in online versions.

3.   The American Association of Cost Engineers: Before going international, the American Association of Cost Engineers was previously created by a group of like-minded individuals who are specialists in the field of Project Management. They are mainly concerned with planning, executing, cost estimating,

and so much more. Currently, it is one of the most powerful bodies in the world of Project Management.

4.   The Critical Path: Ever heard of this outstanding technique? I'm sure you haven't. As a beginner, this type of technique will sound foreign to your ears and that is very normal. As you begin to move from the beginner stage to being an expert, you would now start getting familiar with such technique. Now, what does the Critical Path does? It is an effective technique used to measure the time frame of a project.

In case you want to know how long a particular project would take so as to adequately prepare for it, then this is the right technique for you. As developed by Dupont in 1957, the technique would examine the sequence of activities that has the lowest level of scheduling flexibility. That way, you would get an appropriate time frame.

5.   The International Project Management Association: As one of the world earliest Project Management associations, the International Project Management Association was created in 1965 in Vienna. When it was first created, it was with the motive of allowing Project Managers all around the globe to connect and share ideas under one umbrella. Right now, the International Project Management Association consists of 50 national and internationally recognized Project Management associations and organizations. It had also moved from Vienna to Switzerland and has over 150,000 worthy members from all over the globe.

In order to understand a particular subject matter, it is important for us to first understand its history. History is the vehicle that transports today and tomorrow. Thus, the need for us to revisit the beginnings of Project Management and how it was conceived. This would further shed more light on the concept of Project Management. Be that as it may, this book will take you even deeper. It will help connect your everyday life with Project Management and how you can further improve it with this concept. Want to find out about this? Then follow us to the next chapters.

Chapter Two

Project Management (Definitions and Concepts)

Project Management is a trending topic in recent times with a whole lot of people delving into this line of discipline. Little by little, it had become significant in both our professional world and our personal affairs. Just as the name entails, Project Management simply deals with the art of managing projects as they come. When you successfully handle projects to their efficiently and effectively, you are said to be a good Project Manager. It doesn't necessarily mean one would have a degree or certificate in it before qualifying for the position.

It is pertinent to know that some people have these outstanding skills inherent in them. To them, it's more like inborn skills which they got from birth. Even without learning, reading, or even practicing Project Management, they find themselves extremely good at it. Yes, such people do exist. Nevertheless, if you don't find yourself in such a category, then I guess this is why you are here in the first place. This is why you are picking up this book, so as to brush your crude skills and elevate you from being a beginner to an expert.

Over the past many decades, Project Management had been defined by lots of scholars of the field or discipline and each definition are popularly acceptable. This chapter would help familiarize you with these concise definitions and concept, so as to give you a good idea of what Project Management really is. So the question really is, what does Project Management really entail?

But before we do that, it is important that we know what Project is all about. We can't possibly delve into Project Management without knowing what a Project is all about. When you call yourself a Manager, then there must certainly be something you are managing. This is practically called a project. Anything that has a beginning and an end are a Project. A Project doesn't really have to be something tangible or even official.

A Project is a well-collected effort which consists of basically different parts of a group directed towards a particular goal. According to the PMBOK (Project Management Body of Knowledge) 3rd edition;

"A project is defined as a temporary endeavor with a beginning and an end and it must be used to create a unique product, service or result. Further, it is progressively elaborated. What this definition of a project means is that projects are those activities that cannot go on indefinitely and must have a defined purpose."

With that being said, we can now discuss what Project Management really entails. We aren't farfetched from the truth if we say that Project Management is all about the combination of both Project and Management. Project Management is far more than just heading a group of like-minded individuals. It involves tact, commitment, and perseverance.

According to Jack Meredith, Samuel MantelJack R. Meredith, and Samuel J. Mantel, Jr, I quote;

"Project management is in terms of producing project outcomes within the three objectives of cost, schedule, and specifications.

Project managers are then expected to develop and execute a project plan that meets cost, schedule, and specification parameters. According to this view, project management is the application of everything a project manager does to meet these parameters. This approach to defining project management shares PMI's focus on the project outcomes in terms of requirements."

According to this definition, Project Management goes beyond directing and planning. It also involves focusing on the cost benefits and effectiveness. One of the qualities of a good Project Manager is the ability to make good use of the available resources, no matter how little it might be. Additionally, the ability of a manager to make sure everything goes according to plan is also paramount.

In another view, Investopedia views Project Management differently. In their defense, they see Project Management as something unique. That way, they see;

"Project management as the planning and organization of a company's resources to move a specific task, event, or duty towards completion. It can involve a one-time project or an ongoing activity, and resources managed include personnel, finances, technology, and intellectual property. Every project usually has a budget and a time frame. Project management keeps everything moving smoothly, on time, and on budget. That means when the planned time frame is coming to an end, the project manager may keep all the team members working on the project to finish on schedule."

According to Desmond Cook (The Nature of Project Management, Working Paper, Ohio State University, 1968), he summarizes the definitions of Baumgartner, Cleland, and Gaddis in terms of the project manager's role;

"To produce a product by integrating professional persons into a team operating within time, cost, and performance parameters with that team operating within some lines of organizational responsibilities and authority." Cook goes on to say that projects have four characteristics. They have a single objective, are usually complex in nature, consist of a series of unique tasks, and are normally a one-of-a-kind or non-repetitive activity."

Be that as it may, I believe you now have your own concise meaning of Project Management. Project Management is not just any activity or endeavor, it far more than that. As a matter of fact, it's a strategic process of organizing, planning, scheduling, and even directed to the achievement of one goal – success.

When you see those topnotch companies out there making waves and coming out with a whole new idea, then it's definitely because of their advance and outstanding Project Management team. A weak Project Management team in a company mostly leads to the downfall of that company. Project Management is a very important section of a company. You don't expect the account section to double up as the Project Management team, right? The same way you don't expect the janitors to plan and execute the projects of a company. Thus, the need for Project Management in every company, organization or firm.

Aside from these definitions, this chapter would also look into the important concepts associated with Project Management. On a brief note, we would equip you with these concepts that way, you will be able to fully grasp the real meaning of Project

Management as a whole. These concepts are quite numerous but we will only limit ourselves to the few important concepts of Project Management. They are as follows;

1.   Planning: This is one of the most important concepts in Project Management. Before executing any project, one must first go back to the drawing board and strategize how to go about it. When we plan, we are basically going to achieve success no matter how shrewd the situation or project might turn out to be. Planning is all about mapping out, drawing out, and giving cognizance to a strategic position that would enable you to achieve success in your endeavors.

All scholars of this discipline (Project Management) had reached an agreement on the important role planning plays in managing a project. Imagine a Manager going about directing and executing projects without even a plan? That would be disastrous, isn't it? Additionally, for one to be a good leader, he or she must be very good at planning. When you have everything mapped out already, people will be obliged to follow you, especially when they know there is a well-detailed plan that would enhance their success.

Additionally, planning in Project Management is what had shot up the topnotch organizations of the world today. Even in our personal lives, when we plan ourselves adequately and appropriately, we would realize that things would start taking shape in our lives. We would realize that everything will start looking up and when they don't, there will be no cause for alarm because we must have already made a plan for such an occasion.

We would discuss the term Project Planning further in the course of this book. That way, we would be able to touch every vital part of planning in Project Management. Be that as it may, planning is a concept we can't do without in this line of discipline. As a matter of fact, when we hear the term Project Management, the next thing that comes to our mind is planning. In order words, how we can use planning in making sure our projects are effectively managed.

2. Scheduling: This is also another great concept as regards Project Management. One of the outstanding qualities that make a very good Project Manager is the ability to tell correct schedules and also to keep to it. If we can uphold our Project Schedules, then we would be able to maintain and minimize the use of resources available for that project. In Project Management, Scheduling is a key concept that was developed and aided with the Gantt Chart and the Critical Path.

It became so important that a lot of Project Managers out there would rather stick focus their lens of the concept in managing their project so as to minimize the resources to be used and also to maximize profits that will be accrued from it. Scheduling gained lots of attention in Project Management when Dupont made a profit of over a million dollars after following the Critical Path technique in predicting correct schedules of his projects. This saved him from unnecessary spending, planning, and so much more.

3. Management: Project Management is all about being a good manager. When you have the ability to make good use of the resources available no matter how little, towards the success of a project, then you can parade yourself as a good Project Manager. As a good manager, the first thing you should be able to do when you are faced with a project is to make sure things everything is in order. And when I mean everything, I'm

talking about the machinery, the resources, the manpower, and so much more.

As a concept of Project Management, the ability to manage a project to its desired outcome or destination lies solely with the capability of the management. If the management is weak, poor, and unorganized, then the project would definitely fall apart. In order words, everything that was set to be achieved would not be entirely possible. Be that as it may, in as much as the management is an important concept in Project Management, other concepts are equally as important. One wouldn't be farfetched from the truth if he or she ends up saying these concepts are what makes up a successful Project Manager.

Furthermore, if you want to become successful in this field or discipline, then you need to know the right set of people that would benefit and move your project forward in no time. As a manager, it is your responsibility to hire the right set of people for the job. There should be no emotional attachment when managing a project. Everything should be strictly professional. Thus, good management begets a good project outcome.

4.   Control: If you are a Project Manager or vast in the area of Project Management, then you should know that Control is a concept you shouldn't joke with. In the Old Chinese Empire, the head of their projects wield amazing and overwhelming power, that way, he was able to control others other him and get the job done in no time. If the builders of the Great Wall of China weren't under the command of a single head (Project Manager), then what can be said to be a wonder of the world today would have existed.

Now, that is the power of Control in Project Management. It is not just a concept in Project Management but unarguably one of the outstanding qualities a Project Manager must possess. Make sure you are in total control of your project. In the end, the only person to bear the brunt of the outcome either good or bad will be just you. Thus, having control of your team or project is of utmost importance.

Project Management definition and concepts would further sensitize you with the topic. You will be shocked to know that what you know as the definition of Project Management is a very wrong definition. Things like that happen every day of our lives. With every day that passes, we will learn new things. Thus, I believe this chapter had taught you something new. However, you haven't seen anything yet. Turn the page over and enjoy our next chapter as it promises to be more interesting.

Chapter Three

Project Management (How to begin)

In our previous chapters, we looked into the historical origin (pre and post-historical days) of Project Management and also highlighted the important concepts associated with the discipline. This is solely to keep you abreast and enlightened on the subject matter before finally delving into it. We can't really learn about something new without fully understanding its history. In other words, we can't fully know about something without tracing its root. This is because knowing the history would only give us a better understanding of what that thing really is.

Now, as a beginner, you might be wondering how to go about becoming a good Project Manager overnight. You might also be wondering how much difficulty one must cross before becoming an expert in this discipline. Well sorry to burst your bubbles, Project Management is not really as difficult as others have painted it. It is in fact very easy and can be easily understood if only you would set your mind at excelling in it. Remember, we can only become who and what we want to be if only we learn to believe in ourselves.

Ask yourself these important questions. How did Bill Gates become one of the richest men alive? If Project Management is so difficult as many had painted it, how then did some people get to possess the trait even without learning or studying it? Funny right? If some of us don't even get to drop a sweat before becoming an exceptional Project Manager, why then would you imagine yourself flopping at it? This chapter would enlighten you on the possible ways to start becoming a good Project Manager. It would show you how to start learning Project Management and equipping yourself with the amazing trait in no time.

Trust me, Project Management might sound sophisticated, it is one of the simplest things you would ever learn. Having a clear understanding of Project Management would require the possession of certain qualities and skills. For beginners, you need to first make up your mind and reach a decision on Project Management itself. How badly do you want to be a Project Manager? Are you willing to make compromises and necessary sacrifices when the need arises? Are you willing to learn patiently and attentively? If your answer to these questions is a yes, then there is nothing stopping you from becoming a good Project Manager.

Project Management goes beyond being the head of management or team. In fact, it entails something even more. It encompasses the abilities to manage schedules and deadlines appropriately. It focuses on having a perfect understanding of what the project really entails and possessing the skills to encourage, motivate, and lift the morale of your fellow teammates. Mind you, a good Project Manager must have the ability to sing the praises of his or her team members when the need arises. In order words, you should be able to build a good working relationship with your team.

Additionally, your analytical and budget making skills must be topnotch. When you learn the art of Project Management, these are the likely qualities you would eventually possess at the end of the day. Quite amazing, isn't it? Imagine having all these abilities combine in your perfect little body, you will be formidable at work. You will be able to achieve progress even without trying hard and above all, people will definitely like to associate themselves with you.

Be that as it may, it is important to know that you being the head of a project (Project Manager) don't necessarily make you the ultimate success of that project. It doesn't mean you alone hold the keys to the success of the project. A lot of us out there end up allowing our rate of success gets to our head thereby, we will start thinking the team can't do without us. Remember, no one is irreplaceable. Sooner or later, a perfect replacement for you will definitely come by. Thus, stay humble. Humility is one of the outstanding traits a Project Manager must possess.

Now that you have made up your mind which is the first step, the next thing you should do is to decide on which way you want to go about being an expert in the field of Project Management. It is important to know that there are many ways in which one can end up being a good Project Manager. Like we had mentioned in our first chapter, Project Management is a trending subject matter that had gained momentum over the last few decades.

Thus, there had been lots of amazing and outstanding efforts by lots of people to perfect and reshape it knowledge acquisition process, thereby, making it easy for people to learn, practice, and understand the art of Project Management. Having said that, there is more than one pathway in which one can become an amazing Project Manager. It would interest you to know that all these means are quite reliable and trusted.

Becoming a Project Manager solely lies on your level of reasoning. At one point in our lives, we all wanted to be one thing or the other. At the age of 10, I wanted to become a doctor. At 15, I saw an aviation-related movie at the cinema and wanted to become a pilot. Now, I'm an influencer, a poet, and a writer. Funny right? Being a writer doesn't stop me from yearning to be a Project Manager. If it doesn't stop me, then I believe it should definitely not be a problem for you either, if truly you want to be a Project Manager. It all depends on how badly we want to be involved in the field of Project Management.

Now, what are these possible ways of becoming a good Project Manager? I'll tell you! First, you can learn the tricks and art of Project Management via extensive self-study. Instead of paying a tutor to take you on the subject matter, how about getting the necessary books and start reading? If you read occasionally about anything, then it's only a matter of time before we get extremely good at that thing. If you read occasionally about Project Management, you will definitely be as good as someone who pays to be tutored. That is if you are not even better.

I once wrote a book on the stock market some time ago. But before writing this book, I made sure I read lots of books about the subject matter – Stock Market. It got to a point that I became so vast, good, and knowledgeable about the old and recent trends of the Stock Market. If you don't know any better, you would say I'm a Stock Broker or I own stock. This is the same outcome you would get when you read extensively about Project Management. You will become your own self-qualified and self-trained Project Manager.

Secondly, you can also choose to study Project Management as a discipline in the varsity. This is, in fact, a very good path to follow as regards learning and becoming a good Project Manager. There are lots of advantages attached to this method of learning Project Management. Someone who studies Project Management tends to be much more vast and knowledgeable than someone who depends on self-study. For example, there might be certain terminologies and concepts which can be quite ambiguous and needed to be explained before assimilating, someone who is studying Project Management as a discipline would have the opportunity to meet with his or her tutor for an explanation while someone who is into self-study might not have the same opportunity.

Additionally, at the end of the study comes certification. Studying Project Management in the varsity comes with this benefit. You will be automatically certified at the end of your study, unlike an extensive self-study where there is no certification whatsoever. Thirdly, some of us might need Project Management as additional experience and certification needed in the line of our jobs. This would serve as an added advantage and shoot us up even higher when being paired with our mates.

Thus, all we need to do is to get an institution or establishment where they offer short time classes on Project Management with a certification at the end of the session. This is another pathway in which we can follow in becoming a good Project Management. When you find yourself in a position where many of the qualities and responsibilities resemble that of a Project Manager, then I believe you know what to do.

Nevertheless, the pathway towards becoming a good Project Manager can be quite difficult for some people and sometimes, it can get very easy too for others. It all depends on a number of factors and how well we play our part in this exciting journey. For example, you don't expect to have a stress-free journey into the field of Project Management if you don't even have time to either read or attend Project Management classes. As simple as I might have painted Project Management at the beginning of this book, if you don't take it seriously, then I'm afraid the road towards achieving success in it wouldn't be as easy as it may be.

You need to be serious with it. You need to make sure you truly and perfectly understand Project Management in its raw form. If Project Management is just about heading a group of people, then every Tom, Dick, and Harry would have been a great Project Manager today. You can choose to make your journey towards becoming an exceptional Project Manager a very long one or just a few steps away. It all lies on your hands completely.

Be that as it may, the next process left for you is to take the bull by the horn. Now that you have made up your mind and also selected how to go about becoming a good Project Manager, then I'll suggest you just get to it without wasting your time any further. Mind you, Project Management might sound easy but I never said you won't experience a setback on this exciting journey. Have you considered the financial aspect of it? Have you considered the time frame? Have you considered other temporary setbacks?

If you are faced with financial challenges, then I'll suggest you stick to the first pathway we mentioned above – Self-Study. With technological improvement here and there, you can now download book related to Project Management on your phones and other gadgets without paying a dime in return. That way, you will be able to keep getting more knowledge on Project Management without even paying for it. This is the safest and cheapest way to go about becoming a good Project Manager.

And as regards the time frame challenges, you really need to be certain if you have time on your hands. I have seen lots of cases where people end up registering for Project Management classes and end up getting preoccupied with work and other activities in return. This happens all the time, thus, we need to be certain before entering into it. As the saying goes, always look before you leap. In the absence of all these challenges you might face, then I'll suggest you just Do It!

But before actually doing it, it is important to know that there are two types of certification as regards Project Management. We have the Certified Associate in Project Management (CAPM) and also the Project Management Professional (PMP) certification. Now, the choice is entirely yours on which type of certification you would like to acquire at the end of the day. Additionally, both certification is also awarded and offered as a course by the almighty Project Management Institute (PMI).

I know what your questions would be next after reading through these few pines above. What is the difference, right? I knew that would definitely cross our mind. However, asking questions about what you don't know is a very good thing entirely. As the saying goes, he who asks for directions never loses his way! Thanks to Google Map, asking for directions are quite outdated (laughs), just kidding. Let's get back to it, shall we?

Remember we discussed why people acquire this Project Management knowledge and skills above. If you still recall, I made an example of a working individual that needed Project

Management skills as part of his new functions and also an individual that just wants to study it with lots of time on his side. That is the difference between both certifications. Where one is strictly for those that have more than enough time to spare for reading, classes, experiences, and so much more, the other is just for those willing to get the certificate without going through much stress.

In order words, where PMP requires lots of prerequisites that would give you a hard time and perfectly mold you into one of the best Project Managers out there, the CAPM certification is also there to give you just the accreditation and certification you badly needed without getting to put in much work. Although both of them require you to take classes, read the books, and finally sit for an exam at the end of the session. I believe you now know the difference between both.

Time is the key factor that differentiates both certifications. Now how do you qualify for these different certifications? It's pretty simple. For the PMP certification, there are basically two types of prerequisites which we can choose from. For advanced learners, you can go for the first type of prerequisite which needs at least more than 4,000 hours of working experience and a degree. The other type of prerequisite is for those of a lower set. If you have a secondary education and at least 4 good years of working experience, then you are good to go.

By now, I believe you should know your strength and weaknesses. I believe you should now be able to choose which prerequisite fits your time frame and qualification. That way, you will be able to choose which certification you would want to pursue. There are examples of people who want the PMP certificate but aren't qualified for it, thus, opted for the CAPM which fits their range and specification at the moment. If you fall into this category, then doing that isn't something bad.

You can, in fact, own both certifications if you want. There is absolutely nothing stopping you from doing that. If you aren't qualified for the PMP maybe because of your limited working experience or your qualifications, then I will suggest you go for the lesser category. Start from there. When you attain that certificate, then you can move to the next. It would even give you more experience than any other person out there. It would be like combining both experiences and certificate, which would definitely be an advantage to you in the long run. Project Management is all about leaving the project in the hands of someone capable. Now, what other capable hands can beat yours?

Lastly, maintaining certification comes next. As a certified owner of the CAPM certificate, there are certain exams you would have to sit for every five years or so, just to maintain the certificate. Additionally, it is important to know that these exams can take any form as they change with time, thus, we would advise you to really prepare yourself before going into the exam hall.

And for PMP certificate holders, the maintenance of their certificates takes a whole new dimension entirely. Instead of writing an exam like the CAPM, you will have to make sure you must have completed at least 60 professional development units (PDUs) year in, year out. Now, the question is how do you go about earning these units? It's simple but quite hard at the same time. All you need to do is to make sure you stay active in the world of Project Management or something related to it by running other related courses, either online or in person, going out there to give presentations, dishing out your Project Management skills either as a volunteer or on contract, and so much more.

The real reason for the creation of these PDUs is for you to keep developing and progressing as a Project Manager. I believe if you continue to focus on your Project Management skills without being stagnant, then there will be room for improvement every now and then. That is all you need to know before starting out your Project Management career. The rest is for you to go out there and take the world by storm. Always find the time to learn something new. Also, don't forget to incorporate your experience with everyday endeavors.

Remember, Project Management is not just for our professional use but also our personal lives. You can borrow some of the skills and arrange your life in an orderly manner. You can even help your family and friends out in a time of need. You can help them take control whenever you feel they had lost it. As a Project Management expert, even human beings can be a project in the long run. They can be managed effectively and efficiently.

Now, this is the short and long run of how to go about becoming a Project Manager. And if you are already a Project Manager, then you can still improve yourself with this book. Remember, no knowledge is a waste. No matter how vast you think you might be, there are still certain things you don't know and this book would gladly turn your attention to it. Thus, enjoy every bit of it.

Chapter Four

The Project Manager

The Project Manager is a part of the whole that makes the team. A Project Manager is someone that makes the team functions appropriately and adequately. Everything around us is a project, it all depends on how we view it. Be it a personal affair or a professional one, we all need to be tactful and skillful in order to be able to manage that affair effectively and efficiently. For example, in a family situation where an even comes up and the whole family is expected to come together in order to successfully organize something amazing, someone needs to come out as the head of that organizing committee. Tell me, what better person would be perfect for the job other than someone with experience and certification in Project Management?

You will be shocked at how everyone's eyes will suddenly shift to your direction when the need arises. There would be a need for someone to be in charge and that responsibility will definitely rest on your shoulders. That is the power of a Project Manager. A Project Manager is someone who possesses the ability, skill, and experience to successfully motivate, oversee, and control a certain project to its desired destination – that is its success.

A Project Manager position doesn't necessarily mean someone must possess the CAPM and PMP certification. It also doesn't mean someone must be bossy and too confident. As a matter of fact, a good Project Manager should know that he exists for the group, just as the group exists for him or her. In as much as holding a Project Management certificate writes your name in paint in the field of Project Management, not holding one doesn't make you less of a Project Manager either.

It is important to know that a good Project Manager carries the team along. When the Old Egypt built huge and enormous pyramids for their Pharaohs, it wasn't an individual effort. Instead, it was a collective effort which comprises of the bottom-top hierarchy. The whole chain of command (from the Project Managers to the menial workers) had to digest every possible idea and details together and even relate these ideas amongst themselves. Without this necessary cooperation between the team, who knows if the great pyramid might have collapsed on their head while building it?

Also, in some instances, a Project Manager needs to be confident and sometimes resolute in his or her decisions. If a Project Manager doesn't have a mind of his or her own, then there will be no control in the team. On the other hand, control is an important concept a Project Management. It is a concept the Project Manager cannot afford to lose. Always learn to stand on your words and be confident if you are right or things aren't going your way. You should also cultivate the mind of welcoming new exciting ideas from your team. Your team is like your family, treat them as such.

Categorically, we can boldly say that a Project Manager is a professional in the world of Project Management. At least tons of them have a certificate to show for it. As a professional in a particular field, you are joined with the responsibilities and functions of making good and suitable plans, making procurement of any kind, executing a project to the best of your abilities while following a specific and already designed scope and schedule. If you have any problems with mismanagement and other discrepancies in your organization, then before it gets out of hand, I'll recommend you give Project Managers a chance to fix the mess.

They are not only good at making things flourish, but they are also great at making sure everything falls into place without exceeding the time frame. This is their specialty and what they are trained for. You won't find a Project Manager doing the heavy task in a project, neither will you find him participate fully in the running of the organization. Rather, he will be at the forefront, watching every move with his keen eyes and at the same time making the necessary corrections if the need arises. The Project Manager would oversee the whole operation and steer it to success.

And in cases where the project had attained success and progress, the Project Manager would help improve and even maintain this progress even better. The Project Manager will also strive to keep the mutual interaction between various parties in his team. A peaceful Project team will definitely equate to a high chance of successful outcome. That way, the rate at which the project will fail will be extremely low and there would be a maximization of profits, costs, and benefits.

It is also important to know that the praise gotten from the success of a project would mainly go to the Project Manager and the brunt will also be felt by the Project Manager also. It goes both ways. He or she is entirely responsible for the success or failure of the project. Little wonder why some Project Manager goes the extra length in making sure their projects are indeed a success.

Damian used to be a Project Manager for 30 years before he retired and decides to spend the rest of his life with his lovely wife, Beatrice. In an interview with Damian, he continuously stresses out the fact that during his Project Management days, he would go extra length anytime he feels his project is in jeopardy. According to him, sometimes, he will even stay late in the office trying to crack one or two challenges before the whole team reassembles in the morning.

This is one of the attributes of a Project Manager. Are you getting cold fit already about Project Management? Not to worry, sometimes, the work doesn't even require much stress at all. Sometimes as a Project Manager, you can be very lucky in meeting a group of like-minded individuals in the team. This would even make your work easy. There is also a chance of you meeting difficult people in the course of this work. Trust me, some people can really get on your nerves, especially when they feel threatened or invaded.

Some people can be like that. As a Project Manager, if you are assigned to a new team, there is every tendency that the people of the team would not really be happy and receptive in the beginning. Some might even want you to win their trust. As a Project Manager, you will be taught all this. Thus, you will be more than equipped to face such situations. When that happens, take a deep breath and make sure you take absolute control. Remember, you are the boss.

Additionally, a Project Manager can also double up as the client representative when the need arises. He or she can take action or act on behalf of the client if the occasion warrants it. As a client, if you are not available to take action or decision on a project, then be rest assured that the Project Manager is able and capable of taking appropriate decisions. It is an expertise that is expected of the Project Manager. He or she is expected to take different forms and adapt to the wishes and procedures of the client.

That way, he or she would be able to keep the client happy no matter how difficult the task may be. The goal of every Project Manager is to make sure that the client stays happy no matter what. Thus, he or she should always be up to date as regards the organization's cost, the benefits, the risks, and client satisfaction.

Types of Project Managers

1.   Construction Project Manager: This is a very intriguing section of the field of Project Management. It is a section of Project Management that deals solely with construction projects. If you are having issues with the putting together a formidable team to oversee and execute your construction projects, then hiring a professional in the field of Construction Project Management is your best bet

Currently, there is a valid certification and qualification that had been ruling the world of the American construction industry. Not just anyone can wake up today and claim to be vast in the area of construction project management. Each state also has its own specifications and requirements for becoming one. Be that as it may, a lot of competent people had been disqualified or even barricaded from getting a license, thus, there had been many agitations as regards this unfair treatment.

As a direct response to it, many groups and trade unions have come up with commonly acceptable grounds for all competent construction project management. That way, tests, and exams can be set to determine that. Afterward, the CCM certificate would be issued to you by the Construction Management Association of America (CMAA). That way, you would be able to hold competency and even gain experience.

Additionally, you can now study Construction Management as a degree program at the University. Additionally, there is also a recent development in the colleges today as they now offer Project Management as a master degree course. If you know you already have working experience as regards Project Management, then taking these courses won't be a bad idea. If you have a construction problem, let the Project Managers take care of it.

2.   Architectural Project Manager: This type of Project Managers deal with architectural problems. They are quite vast in the field of architecture. This particular type of Project Management share similarities with the construction Project Management and mostly work hand in hand in making sure the project gets to its final destination and outcome. While the Construction project managers deal with the hard part of the job, Architectural project managers deal with the soft part of the job.

They make sure they focus on the design of the whole project and head the design team. They make an effort in making sure the project goes smoothly and perfectly. They also deal with the issue of budget creation scheduling of the project, and most importantly quality control. These are the core responsibilities of Architectural project managers.

3.   Insurance Claim Project Manager: These types of Project Managers deals mainly with the Insurance Industry. They are always at the forefront of keeping the client satisfied. They are

extremely good at making sure the client gets the best out of their unfortunate circumstances and situations. They are always in charge of the restorations of the client loss, no matter how much.

4. Engineering Project Manager: Do you know that there are Project Managers that specializes in the engineering field? They work closely with other departments in the engineering field in order to come out with the right outcome. They will put together a formidable team and oversee the progress closely. If need be, they can be in the field to oversee things themselves. They can as well work hand in hand with different professionals.

5. Software Project Manager: a software project manager is just as important as any other type of project managers out there. Instead of the traditional way of seeing Project Management (Construction and Manufacturing), software management is a whole new phase which is as important and legit as other types of Project Management we have. Just as other Project Management has extensive knowledge in their field, so it the software project managers.

They have a vast amount of knowledge in software development. A lot of them even hold a formal background in the field of computer science, information technology, and so much more. Software Project Managers are highly skilled and extremely amazing at what they do. They are particularly skilled in the field of software Management. If you have a problem with your software or you have a project at hand that needs utmost perfection, then a software manager is your best bet.

Any good and competent software manager is expected to be vast in the area of Software Development Life Cycle (SDLC). Mind you, having this knowledge doesn't come easy. It requires a lot of competence, skill, and knowledge to attain this feat. A good software Project Manager must have at least a PMP certification.

Responsibilities and Duties of a Project Manager

The Project Manager is like the Grand Master of the whole project. He or she is the brain behind the whole project. He or she is the head that makes sure the body works perfectly. Have you ever seen a train without an engine? Or a car without steering? This is who a Project Manager is. When other teammates start slacking off, it is the duty of the Project Manager to shake them up. However, the different type of Project Managers we mentioned above has their own specific functions and duties. Thus, here are the responsibilities of a Project Manager;

1.   A Project Manager is faced with the responsibility of coming up with great Project Plans.

2.   They also manage project stakeholders.

3.   It is their duty to enable good communication amongst the team.

4.   They are also faced with the responsibility of managing the whole project team.

5.   They also make sure the risks are managed effectively.

6.   Project Managers come up with an amazing budget.

7.   They try to meet up with the schedule.

8.   How about managing conflicts in the team? That is their responsibility.

9.   They motivate their other team members.

10.   They are faced with Project delivery.

The Project Manager is someone who is extremely intelligent and has at least the slightest idea about every department in a project. From the planning down to the execution, he or she must be actively involved in order to drive the project home. Thus, if you want to be this type of Project Manager, the will implore you to read this chapter over and over again. The next chapter would focus on tips to being a successful Project Manager, you don't want to miss it.

Chapter Five

How to be a Successful Project Manager

I'm sure you are smiling and grinning from teeth to teeth after knowing the amazing and outstanding qualities you would hold when you eventually move from being a beginner to an expert in the world of Project Management. A Project Manager is the Superman of the team. He is the kind of person the whole team looks at for solutions to challenges and problems. Now, imagine being that person. Having so much control and power over others and at the same time enjoying the benefits and support of the people around you. Trust me, no feeling beats such feeling.

To be a Project Manager is very easy. With the little experience and formal background you have, all you need to do is to enroll yourself in the PMI (Project Management Institute) as a student of Project Management. You will also need to take classes and examinations at the end of every session. If you feel that is a really big step for you, then you can opt for self-study. Have you ever thought about that too? Self-study is a type of learning process where one teaches him or herself without the help of a second person.

In order words, you will have to read and learn about Project Management all by yourself. However, this comes with lots of responsibilities that will need you to have a whole lot of time on your side. And if you are a busy man, then this is what you should do. There is Project Management certification that doesn't really require much of your time. Instead of wasting time in classes or reading large books, all you need to do is take limited classes and sit for the exam afterward. Be that as it may, it is important to know that all these can only give you the necessary certification, experience, and knowledge that you seek in the field of Project Management but will not make you successful automatically in it.

Thus, I will advise you not to confuse yourself in such a way. The successful and topnotch Project Managers you see or may have heard about didn't get to where they are today by just getting certified. Yes, being certified is a big step in becoming successful in this field. It gives legitimacy and competency to your reputation and skill. A lot of clients out there might not be too convinced even if you are extremely good at what you do. Someone of them might even ridicule or not take you seriously at all no matter how hard you try to influence and prove your worth. That is the power of Project Management certificate.

These topnotch Project Managers have their certificates. But aside from this, there are lots of steps that must be taken, lots of sacrifices that must be done, and lots of things that must change about you if you really want to be successful in this field. According to the recent statistics of the total population of Project Managers (both official and unofficial) in the United States of America, the population is known to have exceeded a little over 5,000. Now, imagine yourself as a beginner in the midst of these 5,000 other Project Managers out there, what a competition, yeah?

As scary as that may sound, I want you to know that being a successful Project Manager doesn't have anything to do with it.

Though the rigorous competition for clients, the stiff labor market, and so much more would definitely have an impact either directly or indirectly on your chances of becoming successful in this field. Nevertheless, a lot of these factors lie solely in you. Before we begin, let me ask you an important question. Do you believe in yourself? Do you have faith in yourself? Do you make the decision to be a Project Manager because you want to or because you just felt like it? Is Project Management what you really crave for?

If you can answer these questions correctly and the answers are promising, then trust me, there is absolutely nothing you won't be able to accomplish in the field of Project Management. There would be no obstacle and challenge you wouldn't be able to surpass. There would be no hurdle you wouldn't be able to cross over so as to triumph and in the end become successful. Becoming successful in this field requires such determination. There are a lot of times when your relationship as a Project Manager and your client would really want to go south. At all times, you need to always work with the 3Cs in order to solve the problem – courage, confidence, and control.

In this chapter, we would look into the various steps in which one can follow in order to be a formidable force in this line of Management. We would also delve into ways in which you can be successful in Project Management. The goal of every businessman or woman, entrepreneur, and even the government is to always achieve success in every one of their endeavors. These endeavors are mostly in the form of projects. A businessman might be willing to create a new pitch or even new software for his declining business so as to change the fate of that business. The best person for the job is a Project

Manager who is vast in knowledge and skill as regards that project.

In another example, if the United States of America is willing to construct a whole new building for the homeless or orphans roaming in the streets, the best way to go about it is to award the contract out to a construction company. It is this construction company that will put together a formidable team which will be headed by a Project Manager. You get my drift? These are examples of how important Project Managers are in society. Project Management is our everyday activities. It is in our everyday life. If we look at it closely, even our life is a project and we are the Project Managers. The way we handle it will determine if we are successful Project Managers or very bad Project Managers.

Whether you admit it or not, Project Management has become a very important part of every professionalism put there. From the construction industry down to information technology, textile, arts and craft, education, and so much more. It's like we can't do without keeping our activities straight and the only people that can help us in straightening them out is the Project Managers. Little wonder why the demand for them is on the rise and many people finding themselves in this position even without much experience and knowledge.

Thus, if you find yourself here out of the blue or through churns of events, then I would advise you to pay very close attention to this chapter. Here, I will be showing you how to come out of your shell and conquer the world of Project Management with your in-depth skills. I'm very sure if I ask you to define who a Project Manager is, all you will be able to tell me is that he or she manages projects. That is correct but doesn't really cover what the Project Manager does. What a Project Manager does is quite numerous and for he or she to be successful at his or her job, then he or she must have perfected these numerous duties.

Now, as a beginner, I would advise you to focus on what is real. That is the only way you can triumph over every obstacle. Additionally, you will need to be very patient as climbing the ladder to success in this field is not a day's journey. Though the journey of a thousand mile begins with one step, one has to be very bold before being able to take that bold step. If you are patient enough, continuously trying hard to better yourself, and being consistent, then becoming successful in this field shouldn't be something hard to pull off. Here are the steps to follow in becoming a successful Project Manager;

1.   Be the people's person: This is the key every topnotch Project Managers out there hold dear. As a Project Manager, it is important for you to be charismatic and extremely friendly at all times. Trust me, your work would only become easier when people around you (workers and team members) love you and want to really work with you. Also, as a Project Manager, there is every possibility that your work would take you to different places where you will definitely meet different people.

The first thing you should do is to the first relationship with the people you will be working with. Make sure you strike an understanding and get them to like you if possible. In order words, don't forget that you are only a Project Manager and not a tyrant. Don't expect all of them to follow your lead immediately. Sometimes, you would have to earn their trust first. This happens all the time. Thus, it is your duty as a Project

Manager to make that happen. If you want to be successful in this field, then you need to develop your people's skills.

If your people's skills are great, then you will be able to keep and maintain your relationships with your workers and clients even better. You will also be able to freely communicate amongst one another. A good communication network is necessary for a successful project. Never forget that. Always pay very close attention to the emotions of your team members. This is also very important. If you can keep to that, then your self-confidence would definitely skyrocket.

Sometimes, the only key ingredient we might need for a project to be successful as a Project Manager is understanding our fellow teammates. When we motivate them when the need arises. When we are impartial in solving crises within. When we stay clear of cultural and religious differences amongst our team, it will only boost the respect and love our team members will have for us. That is the only way to have a good outcome. And a good outcome equates to being successful in this field.

2.   Communication is key: As a Project Manager, when you communicate with your team members, then it should definitely be punchy, direct, and full of reasons. Trust me, the way we talk presents us to our fellow workers. If you are the talkative type, then I will advise you to always caution yourself so as not to lose the respect and control of your team. Learn to communicate and address your team like a professional. There should be time for everything.

There should be time for jokes and time to be serious. Always make that certain in your project team. Additionally, always stress the importance of communication amongst yourselves. Trust me, when you do that, it would relax the tempo and mood of your team. They would find it easy to say what is on their mind, be it an idea or a complaint. In cases of task sharing, always be audible and clear enough. Make sure you communicate clearly so as not to cause any further complications.

Communicate with your clients often too. That way, you will be able to know where and how to help out. Communicate with the contractors, with the customers, with the vendors, and even with the stakeholders. When you have a good rapport with these people, you will be able to carry everyone along and at the same time paving way for your career to be more successful.

3.   Pay close attention to the weaknesses and strengths of your team: A good Project Manager is like a Psychologist, they always make sure they read the people around them as well as their environment so as to know how to better serve or relate with them. Do you want to be successful in this field? Then how about becoming a Psychologist? Always pay close attention to the strength and weakness of your team. Know each person's limit so as not to overload them with tasks and functions.

If you are familiar with everyone's strength and weaknesses, you will be able to relate with them better. That way, your working relationship will definitely improve and the project will become successful in no time. Additionally, knowing the strength and weakness of a team puts you at the forefront of the whole project. That way you will be able to make accurate predictions and even use that opportunity to unlock the potentials in your team.

In order to be successful as a Project Manager, it is important to know your whole team like the back of your hand. That way, you will be able to know which team member would be able to drive your project to success. You will be able to know which team member will slack off and probably pull you back. Additionally, you will be able to know when and how to share these tasks amongst your team.

4.   Gain experiences or more knowledge during the course of your numerous projects: As a professional, there is always an opportunity for you to always upgrade and improve yourself by gaining more experience by the day. In Project Management, there is always an avenue where the Project Managers are liable to further their education either through direct learning or an online session and build their portfolio as well as working on numerous projects (either volunteer or freelance).

But whichever you decide, always know that the choice is entirely yours and both are good for your Project Management career. If you choose to further your studies in Project Management, then it's all good. But if you are focused on building your experience in this field, then I will suggest you start with small projects. As a beginner, that is the only way you can move forward and become an expert. Little by little, you can then move into larger jobs and projects. You will be shocked at how much you have developed and improved after a short while.

5.   Always remember that you are just a Project Manager and not a tyrant: Many people that had failed in this line of Management always fall into this trap. According to some of them, being at the top of the food chain gives one so much joy and happiness that it often get into one's head. It takes discipline and a selfless person not to misuse the power. This is why you should always put it at the back of your mind that

you are just a Project Manager and the project is what connects you to your client and team members.

Don't just give orders, learn to carry everyone along. Many Project Managers end up making this mistake. Don't be one of them. You alone can never know it all. That is why you have other team members in the first place. Always carry them along. Give them a listening ear. You will be shocked at the new and exciting idea they will bring forth. That way, your work will even become much easier. But when you shut your teammates up or even neglects their views, they may end up ganging up against you.

There is nothing wrong in stooping low for your teammates, especially if they have a brilliant idea. Also, there is nothing wrong with giving them the floor when they have something to share. This will never take away your leadership neither will it reduce your position. Always strive to be the Project Manager that keeps everyone happy and excited about working. Trust me, when your team is happy, the project will definitely take good shape.

6.   Always believe in yourself and your team: The beginning of every failure is when you lose hope, trust, and belief in yourself and in the people surrounding you. No matter what, always keep the faith. Even if things aren't going so well, never lose hope in your team. Always find a way to motivate them until the very end. This is the work of a leader. There is a difference between setting expectations for your team and when your team also set expectations for you. Always know the difference and balance these two.

Try to get close to your team members. Know them to a great extent. Spend time with them. If you want, you can even spend time outside working hours with them. At least, that will allow you to know them to an extent. That way, you will be able to know their plans and ideas. If you end up knowing this, then you can perfect it together. Afterward, trust in it no matter what. This will even give them enough confidence and boost their morale even more. Believing in them is what they need in order to overwork themselves gladly if the need arises.

Being a Project Manager is easy, but being the best amongst the rest is not a day's job. It's something that must be achieved by making sure every department agrees to your directives and control. All Project Managers want to be successful. Just like all beginners wants to move up the food chain. You alone can be the Project Manager you want to be. All you need to do is just to believe in yourself and lines will fall in pleasant places for you.

Chapter Six

Common Mistakes you would Encounter as a Beginner

As a beginner, you should always have it at the back of your mind that mistakes would surely happen. Whether you plan for it or not, you will definitely make them and sometimes, in more than one way. Now, when this happens, do not work yourself up or even beat yourself about it. Mistakes only signify confidence and perfection. When we make mistakes, then it shows we are beginning to get the concept of whatever we are learning. Also, mistakes are a reminder that would remind us of how much we have developed or improved.

Even the so-called experts and topnotch Project Managers out there had also made lots of mistakes at a certain point in their life. For example, being obnoxious to the whole team can cause failure and that is a mistake you don't want to make. Some also make the mistake of being a tyrant instead of a team player. They often forget that the Project Manager position is almost the same as that of a team member. Thus, making it look as if they are invincible. These and more are the common mistakes many people make in today's world.

Sometimes, we can't even see the truth in front of us when we are so carried away by these mistakes. It's like a blindfold that veils our very own eyes. It is only after we had committed and made the mistake, then our eyes will begin to see the reality of thing for what they truly are. Additionally, mistakes do happen.

You don't expect to have a mistake-free journey into the world of Project Management as a beginner, do you? Especially with the amount of stiff competition out there in the market, you are bound to make a mistake or two along the way.

Now, making these mistakes is not matter but bouncing back from it and rising from your fall is what should hold substance. This is how successful people in this field are made. This is how they are built. Those topnotch Project Managers you envy today have also had their own share of mistakes and downfall, but they never let it weigh them down. As a go-getter, they were always poised with the mindset of going forward no matter what. Let this be your watchword and mantra, you will be shocked at how much you will progress within the shortest time.

This chapter would highlight some of the numerous mistakes common with beginners like you in this field. It would make you feel normal about it and even suggest probable ways to stop yourself from making such mistakes. As a beginner, many people out there that had just stumble upon or find themselves in this part of the professional world often get bullied easily by the big guns in the game, thereby, making them lose balance and focus easily. That way, they would easily make mistakes due to their lack of self-confidence and control.

According to Damian, he once stressed the fact that a lot of all these experienced Project Managers once tried to bully him into submission when he was still a beginner. He stated that he was once a team member when he started out his Project Management career and was very good at it with fresh ideas running through his mind every now and then. On an occasion where things weren't going the right way and the whole project was in jeopardy, he had the right ideas that would put the project back on time in no time. However, his Project Manager was a stumbling block. Now, what did Damian do? Your guess is as good as mine. He took the bull by the horn.

According to him, he marched straight to the client office one Tuesday morning and poured out his well-detailed idea which was welcomed by the client. The client was impressed not just by his brilliance but by his level of confidence and control. These are the key ingredients you need to possess in order to excel in this line of profession. Now, if you look at this case closely, you would realize that there are lots of mistakes and loopholes surrounding it. Also, there are lots of lessons to be learned from it as a beginner in this field of Project Management.

First, the Project Manager made a mistake of not being friendly and accommodating with his or her team members. Always know that they are your family and not your slaves. Treat them right and respect their feelings & emotions. Secondly, as a beginner, don't make the mistake of allowing a superior break you into submission, especially if you are on the right track. Always exude confidence and do the needful if the need arises. Many beginners are bound to be scared and will most likely keep mute even if the solution to save the day lies in their hand. Always be outspoken, especially when you are making sense. Be like Damian and you will achieve success in no time.

Don't be scared to take action as a Project Manager. If at all you are making a mistake, they always listen to the voice of reason. Always listen to the ideas of your team members. Mistakes have no master. No matter your level of certificates and the amount of experience you have gathered over the years, we are still liable to make mistakes. Meanwhile, when all these happen, be fast about getting them back on track. Delay can be very dangerous. No one wants a bad review on their reputation. Mistakes lead to mismanagement and mismanagement leads to failure.

Additionally, all projects can never be the same. Even if they are the same, the people, the clients, the behavior, the skills around you, the directives, and so much more surrounding the project will never be the same no matter how hard we turn it. Thus, you should stop treating every project in the same way. Make sure you understand and know how to handle each project and the team they come with. Most times, you won't have the luxury of selecting your own team. You might just be appointed to head a team of strangers. When that happens, then how would you go about it? Handling projects the same way would certainly put your projects in jeopardy. Now, here are the common mistakes a beginner in the field of Project Management would certainly encounter;

1.   Incorrect scheduling: This can derail and jeopardize a project in no time. When you miscalculate the schedule of the project wrongly, you are only extending the time frame, thereby causing unnecessary and wasteful spending of scarce resources. That way, the costs of the project would certainly skyrocket and that is bad for business. The client would not only go crazy about this mismanagement but also relieve you of your position if you aren't lucky.

Thus, always get your planning and scheduling right. Don't resist the help of others. Remember, your team is there for a reason. Let them help you. As a beginner, you might not be able to get the correct schedules of the whole project. Don't beat yourself about it. This kind of mistake happens all the time.

But the ability to realize this mistake and correct them is what matters.

2.   Taking decisions and actions independently: This is very common with beginners and experts in the field of Project Management. When the whole team realizes their roles and responsibilities, then things will fall into place. The team is there for a reason and that is to help you make the right decisions. Thus, always ensure that you reach a consensus with your team before taking action.

Always make sure you take them along with every one of your decisions and actions. Desist from taking decisions without proper consultations with your team. There are certain things your team members may add or subtract from your deduction. They may even add a useful idea that will benefit the whole project. In the same vein, I will recommend you always hold a general meeting with your team before taking action.

That way, you will be able to know their views. This will also make them accord a great level of respect to you in the end. When you involve them in the decision-making process, it will further improve the relationship between yourself and your team members. Additionally, there would be a sense of accountability, ownership, and dependence in the team. And this is a very good sign of success.

3.   Always dismantle or dissolve complex projects into smaller bits: Always make sure you engage yourself in this particular act as it would only lead to a more successful project outcome. Are you finding it difficult to plan and execute a project? Do you feel stressed out as the project just seems more complex by the day? Then how about breaking it into pieces? How about

managing it in bits instead of the complicated whole? This is one mistake beginners make.

No matter how they try managing a project, it just gets out of hand. Now, instead of them to break this project into bits, they will be trying hard to find a way through. This unsuccessful trial can be time wasting and extremely stressful. Aside from the energy and time, one will waste, one will also waste resources, and many other important things that will make the project a success. Thus, breaking the project into bits seems like the best thing to do.

Trust me, it will be more comfortable to handle and manage. You will now be able to confidently solve every problem facing the project so as to reach the desired outcome. That way, even the impossible can be possible. Now how do you go about this? It's simple. As a beginner, you just can start breaking up projects into bits without the necessary checks and balances. First, you need to understand the project. Make sure you immerse yourself to it and study it closely before breaking it into smaller pieces.

4. Not keeping your priorities straight: Always know what your priorities are and make sure you separate them from the rest. There are lots of cases where many Project Managers out there give priority to the wrong and irrelevant projects, thereby, losing out in the important ones. If you fall into this category, then you really need to start getting it right this time around. If the project you are handling is just too much, then prioritizing them is of utmost importance.

Always communicate the importance of every project to your team members. That way, they will be able to prepare adequately for the project no matter the level of priority. Many projects end up failing because of the low amount of seriousness placed on it. When conditions that matter or things that should be held in high esteem are being held merely, then the project is bound to fail.

5. Not being the people's person: Project Management entails the combination of different segments into a whole. It involves the coming together of more than one person into achieving the desired result. Now, as a Project Manager, don't make the mistake of not being the people's person. The job is not only about managing costs, benefits, schedules, plans, budgets, and so much more but to also manage the people around you. Your relationship with your team would showcase the kind of Project Manager you are.

Thus, keep your team close. Establish a good relationship with them. Always let them know that their efforts are being appreciated. If there is a good working relationship in the team, then success wouldn't be farfetched. But if the team's relationship is not something to write home about, the project is going to drag on for long and eventually fails in the end. There will be wasteful events. Now, how do you grow your relationships with your team?

Always be the understanding boss. Make sure you are familiar with your team's weaknesses and strength. Additionally, be aware of their emotions. That way, you will be able to tell the exact amount of work each person would be able to handle. It shouldn't be all about work. How about taking time to celebrate with the team outside working hours? That is one of the easiest ways of bonding.

6. Always doing things your own way: Your team is there for a reason and that is to help you take the project to the next

level. Instead of letting them out most of the time, why not include them in every of your decision. Cultivate the habit of communicating wife them when the need arises. And when you communicate, make sure you are professional about it. Always be clear and punchy. That way, there would be no mix-up. For example, if you are sharing roles and responsibilities, then always be audible and clear.

Communication is key to every successful project. You should make sure there is a good communication network between you and the team, stakeholders, client, customers, and so much more. This would definitely aid your work even faster. And always make sure you organize a meeting to aid the communication process. Some of your team members might not be much of a talker, so this meeting should be an avenue for them to share their ideas.

7.   Avoiding the use of Project Management tool: The tools are there for a reason, why aren't you using them? They will not only equip you with amazing features that will help you move your project to the next level but also create a stress-free process for you all through the project. The Project Management tools are there to make your project meet its deadline. This would also make you pinpoint vital boosters which can be used in making sure your project stays afloat no matter the pressure either from within or outside.

8.   Adjust if you have to, and be fast about it: Whenever you feel things aren't adding up, then do the needful. Sometimes, most beginners out there just end up sticking to a particular strategy, method, technique, tool, and other enhancements for no reason at all. For you to be extremely successful in this field, then you have to be very dynamic in your ways. You have to

possess the ability to adapt and adjust to the situation on the ground no matter how shrewd it may be. And when you adapt, make sure it's fast, so as not to end up being too late.

Thus, when things start going south, don't just stick to one particular method. Try to tweak them altogether. Don't be scared of failing. I know you might be wondering what would happen if you fail the project. Thus, the reason for being reluctant in trying out new ways and sticking to your traditional ways. Well, all successful Project Managers had also taken a risk at a point in their lives, thus, don't be scared of doing the same.

Additionally, always keep everyone abreast with the situation of things no matter how bad they might end up getting. This would help you set things straight. It will also help create more time for you and your team in adjusting and coming up with solutions to any problem you are facing. Don't be scared to tweak the budget, the schedule, the cost, the benefit, and so much more so long it would help put your project in a good perspective.

Mistakes can happen to anybody no matter how skilled or knowledgeable you might be. What matters is how fast you can turn the mistake into your strength. Now ask yourself how fast can you make your mistakes your strength? Sometimes, all we need is just a single mistake before we can actually summon up the courage to reach greater heights. Be the Project Manager that always wants more. The next chapter promises to be amazing and quite interesting, you don't want to miss it.

Chapter Seven

New Trends in Project Management

Do you know that there are lots of new and exciting trends that had taken over Project Management? Why go through the traditional process of doing things in this line of professionalism when there are new improved versions and ways in which you can make things happen. Project Management is an exciting field where things change every day. What you know as the most sophisticated and easily applicable tool this year might be obsolete the next year to come. That is how things work in this line of professionalism.

Thus, if you really want to make things work for you as a beginner, if you want to always have control over any project you handle, and if you want to be successful in no time, then you need to stay abreast with the recent trend in this field. One mistake beginners make is to stay glued to a particular technique or method in getting things done in Project Management. This is why I will strongly recommend you always go with the flow each time. Always be familiar with every new trend out there. Make sure you follow the world in every direction they will face.

Additionally, the struggle towards taking Project Management to the next level is very real. Project Managers no longer want to stay stagnant and traditional anymore, thus, leading to the development of more advanced tools, techniques, and methods that had taken over the field completely. In the same vein, I will suggest that you join the ride towards this new and exciting adventure. As Project Management evolves with new trends, it is pertinent for you to also meet up.

For example, instead of racking up your head and brain in planning a project manually, there are lots of new ways on which that function can be executed neatly, easily, and fast. Even though you don't have the necessary power and might to exude these new trends, you can still stick to the old means. But, not without making sure you stay updated always about the ongoing trends. That way, you won't feel left out in the excitement and development. No one will have the confidence to call you an archaic Project Manager.

As a beginner, I will advise you to cut your coat according to your size. Not every beginner out there has the privilege of having the power and means to these new trends. Thus, starting traditional won't be a bad idea. You can plan, execute, and make your project successful by following the traditional means. But with time, you should be able to upgrade your methods, tools, and techniques gradually into new and trending ones. This would even make you become better and familiar with both old and new trends.

Like we all know, Project Management cut across all corners and areas of professionalism. From areas of construction down to manufacturing, textile, education, health, finance, transportation, and so much more. Thus, the need for Project Managers in the world today. If Project Managers are in high need in the society today, would it be at your own advantage if you stay afloat in the tides of Project Management? Wouldn't it be at your own advantage if you become familiar and finally employ sophisticated and new trends into your way of planning and executing projects?

A construction Project Manager would definitely treat a project differently from the way a Finance Project manager would treat a project. An Information Technology Project Manager would treat a project differently from the way a Transportation Project Manager would also treat a project. Thus, it is very important to know that these diverse Project Managers vary in their methods and techniques. But still, it doesn't stop them from employing the same new trend in their business.

For you to be a successful Project Manager and for you to move from a beginner stage to an expert one in no time, then you must stay glued to these trends. This chapter would enlighten you on the various trends that had taken over the world of Project Management. Also, it would show you how to use these trends to your own advantage even as a beginner. Like I explained earlier, these trends are changing rapidly. They are not stagnant like its traditional counterpart, thus, there is a need for you to always stay updated.

The year 2019 had been great so far as regards Project Management. There had been lots of exquisite trends that had been introduced in the field of Project Management. So what are these trends? Here they are;

1.   AI (Artificial Intelligence): The world has changed a lot in terms of technology. Thanks to new innovations and development, Artificial Intelligence has come to stay. It won't be farfetched from the truth if we say that AI is the latest trend in every aspect of the world. In every sector and industry, this new trend had found its footing and at the same time become the driving force behind every success. It is the talk of the town.

There are lots of applications which would make your work very easy. For example, Siri on Apple and Samsung devices. This app is a talking AI which would help you put things in order. You will be shocked at how fast Siri would help you organize your project. It would help you collate the figures properly. Google is also a search engine that would make your work faster. In case you are stuck or making no headway, you can easily check Google and find out how many people had solved the problem you are having.

That way, your productivity, and output will definitely triple. AI will give you a platform where you will be able to connect even better with your team members, client, stakeholders, and so much more. With lots of amazing applications to choose from, you will be able to move your Project Management skills to the next level. Your work will be more efficient and effective. You will be surprised at how these amazing applications will give you vital suggestions, correct your mistakes, and arrange your work neatly.

2.   Hybrid Approach: It is important to know that the traditional method that seeks to use the same process and procedure doesn't really hold value any longer. It is pertinent to know that this one-size-fits-all-strategy no longer hold the same power as it had in the past century. In recent time, there are lots of approaches that are now trending and taken the place of this traditional approach.

Additionally, you can also interchange both traditional and modern approach to suit your project. It all depends on the kind of project you are handling. This is what is known as the Hybrid Approach. A lot of Project Managers out there don't just rely on one particular type of approach. Like we said earlier, what makes a successful Project Manager is the ability to tweak these approaches. When the project seems to be more complex and extremely large, then I will recommend you to make use of the hybrid approach.

Project Management today has taken a whole new dimension. Projects are no longer as easy as they used to be. They are now shrewd, complex, and tougher. Thus, the need for a switch between the traditional and modern approach – Hybrid Approach. Be that as it may, I will advise you to be a hybrid Project Manager. That way, you will be able to handle any project that comes your way with any problem.

3.   Emotional Intelligence (EQ): Emotional Intelligence is a new trend in the world of professionalism. Many people don't pay much attention to Emotional Intelligence but trust me, this is one trend that has taken over the world by storm. A lot of companies out there has now given Emotional Intelligence a lot of recognition and cognizance. There is a possibility that we might get lost in the advanced methodologies and techniques as a beginner, thus, the need for being Emotionally Intelligent.

Additionally, Emotional Intelligence would give us a much-needed advantage over others. By understanding other people, we will be able to know how to relate with them as well as influence the decisions of our team members and clients. Emotional Intelligence is a very useful weapon that we can use to reach a greater height in the field of Project Management. Emotional Intelligence will build you up emotionally and your relationship with people would improve rapidly.

This would ensure your career follow a straight path. If your team is filled with people of different cultural, ethnic, and religious background, Emotional Intelligence would help you manage them perfectly. It would help you understand every individual irrespective of their background. You will be able to resolve conflicts in the team. You will be able to understand your team's emotions. Trust me, Emotional Intelligence is the new trend.

4.   PMO (Project Management Office): In recent times, the PMO had been gaining momentum due to their outstanding and amazing performance over the last few years. It might interest you to know that the PMO holds the key to the success of your project. In order words, you cannot have a successful project without paying extra attention to the PMO. In case you are wondering why the PMO is so important or becoming increasingly popular by the day, then here is why.

The PMO is the bridge between project failure and its success. It is the stimulant that ensures all projects get to their desired destination and that is being successful. It is what pushes the project into attaining its goals and objectives without many hurdles along the way. The PMO stands between the client's decision and the Project Manager's execution. I believe you can now see why you need to really hold the PMO dear to your heart as a Project Manager.

Be that as it may, recent research has it that about 42% of the total projects executed in the United States of America were only able to meet their target, goals, and objectives. And this is because they have an indomitable PMO behind them. This fact gives much cognizance and importance to the recent trend of the PMO. It shows they are the driving force of every organization and would make the project easier for you if only you can learn to acknowledge them.

5.   Kanban Boards: Have you ever heard of this particular trend? If you haven't, then don't beat yourself about it. Remember you are still a beginner in the field. Besides, not all Project Management experts in the country can fully beat their chest with the knowledge of this concept. Be that as it may, Kanban is not an English word. It is, in fact, an Asian (Japanese) word meaning "Billboard". The story behind Kanban is a very funny story. Initially, it was an idea introduced by the Toyota organization in Japan to take their manufacturing process to the next level but had now found its way into the realm of Project Management.

What is Kanban? Kanban is an approach which entails the use of different Kanban cards which are spread out on a large visual board for the purpose of managing workflow and the manufacturing process. In order words, it is a way of making sure the manufacturing process are well accounted for and in the process leads to a more improved way of managing the affairs of the organization. For Project Managers, Kanban would serve as a new way of finding new ways towards managing the project.

Instead of sticking to the old methods, Kanban would serve as a whole new method one can make use of. It would also open the minds of Project Managers into seeing things differently. Kanban is indeed a sophisticated innovation. When your projects are flexible and change frequently in time, schedule, and so much more, then this Kanban approach is the best approach to employ. Today, Kanban is a force to reckon with in the world of Project Management. It would allow you to hold control of the project and push stagnation further away.

6.   Analytics: Most times Project Management is all about crunching those numbers. Thus, what better way can one go about it other than Analytics? Analytics is a better method of calculating and dealing with numbers. Instead of going about it the old and traditional way, Analytics will provide you with a more sophisticated way. In addition, it's going to make calculations easy and less stressful. What you would normally spend hours racking your brain on would now be solved at very little time.

Analytics would give you more time to look into the details of your results. That way, you would be able to easily sieve out what is wrong from what is right. Aside from the calculations, you can also check out the performance of the team so far. One advantage of Analytics is the ability to present you with raw facts and data. These are many more are the reasons why it is trending in today's world.

7.   Cybersecurity: It is important to know that the internet comes with its advantages and disadvantages, with its blessings and its curses, and with its strength and weaknesses. No matter how little and insignificant your data might be, you never can tell who is snooping around. If you are a beginner in this field, then you should know that cybersecurity is an important trend which must be taken seriously so as not to lose your data to cybercriminals.

You should know that there is a need for you to always have at least one potent firewall installed in your devices. This would help sieve out hackers that are willing to steal your data. Always secure your plans else they will end up getting compromised with you even having the slightest idea. Every Project Manager has his or her own secret strategy, plan or method that works magic for him or her. Unless you are willing to share this secret weapon of yours, then I will suggest you keep it safe no matter what.

Don't be left behind. Don't be outdated and archaic. Be the modern Project Manager by making yourself stay abreast with different new trends that have taken over the world as a whole. Sometimes, clients like seeing that their Project Managers are sophisticated and modern. This would keep their mind at peace. Keep yourself updated and join the winning team.

Chapter Eight

Project Planning

As a Project Manager, this is, in fact, one of your main jobs in the field or outside the field. The term Project Management alone depicts serious planning, thus, making the Project Manager the best planners amongst planners. We all get to know a good Project Manager from the way he or she manages the team and plan his or her events. Additionally, the success of the project lies solely with the planning skills of the Project Manager and his whole team.

Imagine being a beginner in the field of Project Management and you don't even know how to plan a project effectively and efficiently, what does that make you? You will instantly lose the support of the team as well as that of the people around you. Aside from planning for a project, how about planning our lives appropriately? First, we are an individual before becoming a Project Manager, thus, the need to plan our lives. If our personal affairs are in order, then it's going to reflect well on our professional life.

I'd seen lots of cases where people fail to plan their personal lives well. They would give little cognizance and importance to their lives and end up messing up in their place of work. Our personal lives matter. A lot of people would come to work every day with a sorrowful heart. Sometimes with a very dull mood, thereby, affecting everybody they come in contact with at work. Don't think because you are a Project Manager, thus, that makes you immune to such things. This assumption is very wrong.

As a Project Manager, it would only get worse, trust me. When you lose control of your life, then everything would certainly fall apart in no time. Clients would start to doubt your abilities. Your team members would start to revolt. The projects would not take good shape. Now, that is a bad thing. According to Damian. He mentioned that he had once faced this adversity in his life once or twice and it really weighed him down. His personal life was a mess and it really made him lose control and focus. Planning a project effectively became very difficult.

He mentioned a small tiff with his wife which escalated into a bigger one just because he was a good planner as a Project Manager but didn't know he was a very bad one at home. It took him a lot to realize this fact that was staring deep into his eyes. And when he did, he made sure he never made such mistake ever again. That is the power Planning wields. Now, if this can happen to anyone when making awful plans in your personal life, what do you think would happen to you if you really suck at Project Planning?

I will strongly recommend you become read extensively about this topic. It is one of the most outstanding attributes in a Project Manager Clients would always notice first. In order to move from this beginner stage to an expert one, then you need to up your Project Planning game. Now, what would this chapter do? This chapter would familiarize you with what real Project Planning is without sugarcoating its true meaning. It would immerse you with the fact that Planning is very important as a Project Manager and as an individual. And all these would be alongside the processes and steps to follow in order to plan a good Project.

Be that as it may, you will be shocked to know that Project Planning is more than just sitting down alone in your office while scribbling down diagrams and codes best known to you. It is more than that. It is, in fact, an aspect of team bonding. It is a way of connecting with your team. It is a way of proving the point that "many good heads are actually better than one". This is what planning means. It is a means of drawing up points, ideas, and will in order to execute an action. That is what planning means. Remember, you can't do this alone. Your team is there for a reason, always carry them along. And if you feel you must do this alone or had already perfected a plan before the arrival of your team then feel free to share your amazing plan with them. You will be shocked at how helpful their feedbacks would be.

Just imagine this scenario as a beginner in the field of Project Management. After a careful perusal of your credentials and resume, even after knowing that you are still a newbie in this line of professionalism, the client still decides to take a chance on you. I'm sure you would want to impress his or her that choosing you was a wise decision. Now that you've gotten the job, what do you plan on doing first? Sharing your plans becomes the next best thing to do, yeah?

You would want to blow their minds away with an amazing performance and presentation. In this line of professionalism, the first impression matters a lot. So, let's say you have every necessary tool, technique, the methodology at your disposal, how then do you go about making the best out of them? How do you go about making a very good Project Plan? If your project is quite difficult for you to handle, then breaking it into bits in order to effectively manage it becomes the only solution. However, if it is relatively easy at the same time, then lucky you.

Now, how do you go about your project plan? Remember your client and stakeholders would want to see an impressive presentation, how would you go about it? After making you the Project Manager, they must have kept you abreast the situation of the organization. They must have familiarized you with their hopes and aspirations. They must have shown you what they are expecting from you and their hope is for you to deliver. All these might seem crazy at first. As a beginner, it might seem very hard to comprehend. Trust me, I've also been there.

Especially when you start seeing terms and concepts you've never come across before. Now, this is where you need to get your act together. This is where you need to be focused and remember, do not panic no matter the pressure or complications. Even if the time frame is short, always be confident of yourself. Know how to utilize every tool you've got. Your team is a helping hand, use it well too. There are things they know and you don't, don't be too proud to ask if you don't know. We all don't know everything.

Now, how do we plan a project? Where do we go from here? How do we go about it? What steps do we follow to come out with a good plan? It's pretty simple. This chapter would enlighten you on the six basic steps one can follow is coming up with an amazing plan which would blow the minds of your client and stakeholders away. Here they are;

1.   Understand what your clients, customers, and stakeholders want: This is the first step towards reaching the goals and aspirations of your clients, customers, and stakeholders. You

can't plan a project without knowing and understanding what they want. Remember, the project you are about to plan belongs to them in the first place. Thus, make sure you truly understand where they are coming from. That way, both parties would remain happy at the end of the day.

Don't assume for anybody. If you know you are not clear on a particular goal, go back and ask. Make sure you are clear about what they really want. In the end, they are the ones that will be affected by the end result. Additionally, as a beginner, it would be great if you can jot down these goals from your clients. Make sure you scribble every little detail down. These should be the guiding principles that would guide you through your project.

Immerse yourself with the kind of expectations they would like to see. Use this information to create your project cost, scope, budget, scheduling, and time frame. When you do this, make sure you follow the right channel, thus, making everyone stay on the same page. This would reduce the risk of miscommunication. Also, you can be creative. Instead of settling for what your clients want, you can also tweak some of these goals and make it look more appealing and enticing. This would boost your Project Management profile.

2.   Make sure you prioritize the goals: This is very important so as not to waste more time, especially if the time frame is very short. Always make sure you set a scale of preference for the goals. Sometimes, your clients don't always know what they want. You might be faced with a little challenge of your client listing uncountable goals. And you and I know that all these uncountable goals can't fit into one plan. Thus, the need for prioritizing.

After understanding the goal and objectives of your clients, customers, and stakeholders, then I will suggest you sieve the best out of the rest. I will recommend you set your priorities straight, thereby, making sure your work remains as easy as it can get. Now, it is important to know that these sieved goals must reflect your client's wants and views. If they don't, then you need to start all over again. Make sure they are very clear and easily understood. This would give your team members easy access to the goal, thereby, making the project easy altogether.

3.   Take the time frame seriously: There is no better way of tasking yourself into being productive other than making sure you follow the deadline. When we put the deadlines at the back of our minds, we would be shocked at how effective and efficient our work would be. Ensure you know how to go about it also. Make sure you always know what the deliverables are. You should draw out the amount of time yourself and your team would invest every day into the project.

If you are far behind, then you should know how much time you would invest in the project. Some Project Managers end up not going home all through the night. They sometimes spend more time in the office than at home. It depends on how difficult the project is and how short the time frame is. Setting the time frame to your taste would enable you to know what your progress is from time to time. In order words, you will be able to check how far you have gone.

4.   Come up with the right schedule: Scheduling is very important in Project Planning. One mistake from miscalculations and misguided schedules, then everything about your project is going to take another form. From the

costs, down to the timing, the benefits, the scope, the budget, and so much more. Scheduling allows us to determine the right proportion of time, costs, and so much more to be given to a task. This would also tell you which of your team member would be responsible for the task.

Create a Gantt Chart for your scheduling process if possible. This would only allow you to locate and infuse your dependencies with the task. The Gantt chart is a mechanism that makes scheduling easy. Always make good use of it. Don't forget to share these tasks amongst your team. Remember, each member of your team has an area in which they specialize in. Give the members tasks that are best known to them so as to achieve the best results in the end.

They alone know how to handle the tasks effectively and efficiently. Your job is to make sure each of them comes with a good result afterward. Manage this congregation of brains well. Make sure they are always in line with whatever plans you've made so far. That is the only way you will be able to draw out an amazing plan at the end of the day.

5.   Know the challenges and risks you are facing: The only way you can cross over a hurdle us to identify and know what it is. This is the only way you will know which type of solution to proffer. It is important to know that all projects will definitely hit rock-bottom. This can be either at the beginning, in the middle of it or even towards the end. As a beginner, I will advise you to keep this in mind so as to be readily prepared for what is coming.

Trust me, the risks will come. And when they do, always make sure you have them contained. That way, they won't pose a threat to your plans. Don't just sit there hoping for a miracle when they come knocking. You need to take the necessary steps into making sure they don't escalate into something bigger. You work as a Project Manager is not only about managing projects, it is also about managing conflicts, challenges, and at the same time risks.

You need to make sure you had foreseen any challenges or risks you might face along the line before starting out your planning process. For example, if you are working with a group of culturally, religiously, and racially different individuals, then you need to put it at the back of your mind that there is bound to be a religious, cultural or even racial conflict along the line. Thus, you would know how to effectively manage the team.

Additionally, you can always make sure your projects stay risk-free by conducting a risk assessment. Also, you can develop your own strategy for managing the risks surrounding your project planning process. That way, your project plan would be free (at least if not totally) from emerging risks.

6.   Always share your break-through and update to your clients and stakeholders: This is the problem some client face with lots of Project Managers out there. When they have assigned a project to a Project Manager with a specific deadline, they no longer see that Project Manager again till the day of the presentation. This act is not really advisable, especially as a beginner. Always make sure you send updates to your client. If possible, meet them in person. They might have a new goal or angle they would want you to focus on.

Additionally, always make sure you keep your client abreast with the progress of your work. If you feel you don't want them to know what you are planning before the deadline, then you might really need to stand on your feet in order to make that happen. Some Client might prove very difficult, trust me. They would not want to make it easy on you and sometimes want to exert their authority on you. The best way to stop that from happening is to tell them beforehand. Let them know how you operate and how you like doing your thing.

And when you are making your presentations, remember first impression matters. Make sure your presentation is clear and can be easily understood by all. You can throw in a few Project Management terms to sound official and professional but don't go overboard. What is the need of going all grammatical without passing a message? Break down every complex term in order to pass your message. Trust me, communication is key when making a presentation that would blow the minds of your audience.

Planning a project is quite simple but yet complex. With the right combination of resources, putting a plan together shouldn't be added to pull off. But what if these resources are limited? How are you going to pull it off? As a Project Manager, you are expected to be versatile and this involves working under unfavorable conditions. The Project plan is vital in the success of a project, thus, make sure you pour your attention into it. The next chapter would focus on Project Control. Please flip over.

Chapter Nine

Project Control

This is a very important concept in the field of Project Management. As crazy as this may sound, a project can only be successful if the Project Manager holds enough control. Do you expect a good result from a mismanaged team? Do you expect something good out of a disorganized team? The answer is no. This is why having control of your team is very important. I want you to keep it at the back of your mind that it's not going to be easy having control of your team as a beginner. With just a little experience, many of your team members would not really want to give you their support at first.

Especially if some of them are more experienced than you. There is always going to be a need for them to act strange and most times see you as inferior. Now, when this happens, there is always a need for you to prove yourself to them. There is always a need for you to earn their trust. There is no need for violence neither is there any need for unnecessary tiffs. All you need to do is to reaffirm your position by showing them you are a part of a whole. You need to show them they are equally important and that together you will make the project a success.

Having control of a project as a Project Manager is paramount. Remember, you are in charge of managing the affairs and running of the whole team. Thus, you need to always be in charge. There is no way you would make your team work together in harmony if you don't manage them with understanding and perseverance. Having control doesn't equate to having them fear you and submit to your will even when it is clearly against their better judgment. Sometimes, most Project Managers get it wrong in this aspect. They think having control means having the rest of the team at your fingertip.

This is very wrong. Having control of your team also entails having them enjoy working with you, thereby, making them submit to your will effortlessly. Having control means having the ability to influence the views, decisions, and opinion of your team, the client, and even the stakeholders. That is real control and that is what you should aspire to have. When you have that, then the rest will definitely fall into place. The trust and support from your team will automatically grow even without you giving it your best shot. Be that as it may, not every Project Manager has to go through the struggle of blending with the team.

Some are just natural leaders and possess inborn skills and charisma to make people fall in love with them. In other words, you might also find it easy to connect with your team even without really sweating it out. When you possess these skills and qualities, it would just make your work even easy. When you get your clients, team, and stakeholders to like you as a Project Manager, then you have automatically gained controlled without even knowing it. You will be shocked at how your clients would suddenly start becoming lenient with you. Your team would start becoming receptive and amazingly sweet. That way, making your project achieve the best possible outcome.

Now, what does this chapter tend to do? This chapter would familiarize you with what Project Control really is. It would differentiate and separate compelling with influencing. That way, teaching you the difference between the two before you start confusing both. Lastly, it would tell you how to easily exert, hold, and maintain control as a Project Manager. Trust me, you don't want to find yourself in a situation where no one in your team listens or acknowledges you as the team leader. It really sucks and at the same time would make you look terrible at your job.

But when this happens, don't sweat it. Don't beat yourself about it. Instead, look for possible ways on how to turn things around. Make sure you brace yourself up as a beginner as this may be your fate with any team whatsoever. If you are struggling with control issues on your team, then I will suggest you pay close attention to this book, especially this particular chapter.

Project Control involves a whole churn of procedures and events that involves both the cooperation of the whole team and the authority of the Project Manager throughout the project. When the Project Manager has control, he or she exerts absolute influence over everything that involves the project. From the cost levels to the budget, down to the schedule, and so much more. Now, you begin to ask yourself if the Project Manager is just there to exert his hold over everyone and everything. Well, this is another way to put it. There can be no ship with two captains. The ship would definitely capsize in the end due to disagreements and headlocks. This is the same thing for a project. It will definitely fail in the end.

Thus, the need for Project Control. When you confidently oversee the tasks and levels of each part of Project, then you boldly say that you are in control. Mind you, some client can be very controlling. Remember, it is their project you are dealing with ad aside you, no one wants the best for the project aside from the. To them, the project is more than a project. It is like their child, thus, they would most times get skeptical about every turn and decision you make. Project Control is compulsory traits or skills the Project Manager must infuse in his or her Project Management experiences so as to be able to come up with amazing results.

This is not something he or she should have for the benefit of the project alone. It is something he or she must possess as a Project Manager. If you have control over your project, then there would be no problem of mismanagement. Everything would definitely play out as it was planned exactly. Be that as it may, when we mention the term Project Control, then you should know we are only referring to three important things – Setting Standards for yourself and your project, Measuring Performance all through the project, and at the same time Taking corrections when the need arises. Thus, here are the important ways to making sure you gain control of your team, your project, your client, your customers, and probably your stakeholders.

1. Make sure you always hold meetings: Having control can also mean giving listening ears to your team members. When you respect their opinion and listen to their view about a particular task, then you are in turn creating an avenue for team bonding. One of the most effective ways of doing this is to hold meetings with them regularly and efficiently. In the meetings

make sure you outline and reemphasize on the need for completing the goals and objectives of the project.

Throw in a few friendly jokes if possible. Make the atmosphere mild and conducive for all team members in any way that you can. If you show your team members that you trust and acknowledge their view, they will, in turn, give their loyalty to you in return. That way, you would be controlling them even without trying. Make sure the communication network is very effective. As a Project Manager, the only way to control your team is to manage them appropriately. This is your job. To utilize every resource towards the success of a project. Don't forget that.

Always show that you are a professional in this field, that way, they will be more submissive. Begin every meeting in a good note, just like a professional. These meetings will show you how far you had come and how far you need to go. The meetings would also showcase how much control you have on the team. Trust me, the control process can be very complicated. One minute it would look like you are great at it and the next, it would totally look frustrating. Thus, always be yourself. Share the tasks according to the skills your team members possess. Don't be bias in your dealings. Always make sure you are clear and understood.

2.   Always perform quality control in your team: Additional, you might want to check the amount of hold you have on your project and how much this hold had delivered. The best way to go about this is via Quality Control. This would show you where you need to focus more on in the project and which areas need a little slacking off. Quality control is there to make you exert your influence on the team as a Project Manager. It is a

reminder to your team members that you are in charge, thus, your decision on any task is final.

In a situation where you are faced with insubordination or lack of control in your team. You can check these situations by performing quality control in the department where this insubordination is emanating from. In order words, you can use quality control as a weapon which would shut down this gross misconduct. Quality control would put more pressure on them and more grace to you. Now how do you perform Quality Control? It's simple.

Quality Control is more like a test which is carried out on a product before it finally reaches its final destination – the consumers. Be that as it may, your clients would definitely want to see the quality of the product or project you are heading. Thus, one can say this is a compulsory phase you must pass through. Nevertheless, it is a weapon you should use in order to gain the respect and control you truly need for a successful project.

3.   Always measure and be updated of the progress so far: This is one hell of a weapon that will keep your team on their toes all through the project. When you set timelines and deadlines for a particular task, it shows you possess the authority and control over your team. It also shows that your team is cooperating and that is a good thing. But in moments where you can't even tell them what to do and when to do it, then you need to follow this step.

You should cultivate the spirit of monitoring, staying updated, and measuring the extent to which your project had progressed. This would give you a certain amount of control. Additionally, it would also ensure that your team takes their jobs seriously and there would be no issue of delay. Staying updated is your job but you can also make good use of it as a way of influencing your team. If that happens, then your project would only have one outcome – success.

Additionally, if you feel a certain team member or members are acting strange and causing unnecessary trouble in the team, then feel free to reassign them to another task. Staying updated would give you the much need advantage and ideas you need in making sure you stay in control. You will be able to tell who and who is loyal, who and who are really not happy about the team's condition, and so much more. That way, you would be able to find the right solution to the problems.

4.   Always make sure you are active and responsive to changes: Trust me, a weak Project Manager can never succeed in this line of professionalism and I trust you don't want to fund yourself in that category. Thus, always take the necessary action when the need as rises. Never stay dormant as the leader of the team. This would only lead to insubordination and misconduct. As the head of a project, you should know that you are responsible for the failure and success of the project.

You should be able to know what is wrong from right and also know the appropriate decision and action to take in line with that. This would make sure everything goes in order. It would ensure that the costs, schedule, benefits, budgets, and so much more remain the same. The only constant thing in life is change, thus, the project will definitely take a new course with time. That is why you need to make sure you take the appropriate decisions and be ready when this happens.

If a team member misbehaves, then you should know what to do against that. If a team member performs a task well, you should know how to act also. Additionally, if the project takes a whole new shape (for example, the client wants to focus on a whole new goal and wants you to tweak this with the project without losing the point of the project), then you should be able to also make that happen.

5.   Be a perfect judge amidst conflicts: One way you can use to gain control, support, and respect of your team is to be diplomatic in all things. Your ability to manage issues perfectly would always come in handy, trust me. Remember, you are a manager. Not just Project Management but also Human Management. You should also be an expert in Emotional Intelligence so as to be able to deal with the issues that would definitely come up later.

Additionally, as a Project Manager, you need to manage not just human relations but also the important issues surrounding your project. You would need to make sure everything goes hand in hand, else, the project would definitely fall apart. The cost should complement the time, the timing should complement the benefits, and the benefits should complement the budget. This is the job you've been appointed for. Know these issues, understand them, and come up with an amazing solution in the end.

Don't forget that you can always ask for ideas and opinions from your team members if you are short of it. Your team members are there for a reason and that is to help you out. Trust me, it won't make you smaller or lose control. Instead, it will only show that you are also a team player.

Having control is different from compelling or forcing others to do your bidding. Please don't confuse the two. What I am possibly referring to is when one possesses an influence on your team, client's, shareholders decisions and actions to an extent. This influence can come as a result of love, admiration, agreement or any other source. The difference between influence and compelling is a force. When your team starts fearing you, then you need to go back and check yourself. Ask yourself where you got it all wrong. Your team shouldn't be scared of you, instead, they should hold you in high esteem.

This is the kind of control I am talking about. Be that as it may, if you know you are having issues holding control in your team as a Project Manager, then there must be something holding that control. You need to go back and ask yourself what went wrong, then work towards correcting this flaw. Project control helps in making sure the project gets the desired result. Now, let's talk about the enhancements that would foster the project into reaching its desired destination. And when I say enhancements, I am talking about the key techniques, methodologies, and tools in Project Management. These and more are what the next chapter would delve into. Let's get to it, shall we?

Chapter Ten

Key Techniques, Methodologies, and Tools of Project Management

In Project Management, there are lots of methodologies, techniques, and tools which can be employed to spice up your project. You might be wondering why lots of topnotch Project Managers out there excel in any project they lay their hands on, well it's simple; they basically make use of these diverse enhancements in making sure their projects reach their desired destinations. Whenever they feel the project isn't making any headway, they turn to these diverse techniques, methodologies, and tools for help.

Many people have this notion of Project Management having a one-size-fits-all technique, method, or tool. This is a very wrong notion. With lots of exquisite enhancements out there that would make Project Management easy, fun, and successful, there is none of them that possess such quality. Don't get me wrong, each technique, methods, and tool are very flexible, reliable, and even dependable but none of them is universal. Each of them has a specific function and reason for their creation. Now, it is left for you to select the right one at the right time.

According to the Project Management Institute (PMI), and I quote;

"a methodology is defined as 'a system of practices, techniques, procedures, and rules used by those who work in a discipline."

These enhancements like I call them because they further help enhance Project Management in more than one way are very numerous in number. But for the sake of time and your level of comprehension, we would only mention a few from each part – Techniques, Methodologies, and Tools. Their primary aim is to help the Project Manager out in making sure they achieve success, no matter how little in their projects. Mind you, each of them has their own set of principles and rules. Thus, you alone will be able to decide which one is good for you and which is not. But it all depends on the nature of your project and what problems you are facing at that particular time.

Remember, these enhancements are there to make your work easier, thus, feel free to employ any of them when the need arises. Also, you can choose to use more than one enhancement at the same time. It all depends on the nature of the project you are dealing with. Thus, feel free to change from any when the need arises.

If these topnotch Project Managers can employ these enhancements to the success of their projects, then I would advise you to also pay close attention to it. This chapter would familiarize you with the basic techniques, methodologies, and tools you can employ on your projects. It would also go further to explain how you would make the maximum use of these enhancements. But first, we will begin with the key techniques, before going for the other two.

Key Techniques in Project Management

1.   The traditional Project Management: Aside from the new trends that had ruled the world of Project Management in recent times, traditional Project Management had been a very reliable technique used and followed strictly in order to achieve remarkable results in the end. Here, you just need to perform the traditional roles of a Project Manager. You even carry your team along, share ideas, execute projects, and plan accordingly. If you are working on a less stressful project, then this is the best technique you can ever use.

2.   Waterfall Project Management: This is another way of using the advanced traditional technique. The waterfall Project Management is another technique that requires every team member to focus more on the common goal. Here, there is a need for cooperation. Instead of team members working individually on tasks, they can come together and work closely. The waterfall project requires a whole lot of team members. This team cooperation often leads to team bonding and a better project result. This technique is very dependable on the Gantt chart.

3.   Rational Unified Process: The rational unified process as a technique got its name from its developers. It is quite familiar and shares the same tenets with the iterative style of software development projects. If you are working on a consumer-product kind of Project where the consumers would have to give you their feedback, then this is the technique for you. Unlike the waterfall type of projects, the rational unified process focuses more on the views and opinions of the end users. It believes and works with the consumer view for better improvement.

4.   PERT Project Management: After the cold war, there was a need for a new way of perfecting large scale Project Management. Thus, the result of the collaborative effort with the armed forces is what we know as the PERT Project Management today and it has been very helpful ever since. This particular technique is quite amazing for a one time project, especially in the manufacturing stage. Although there is always a high chance of the project stretching out more than normal. Additionally, if you want to know how far you've come in a project, then this technique would come in handy.

5.   Critical Path Project Management: One can say the Critical Path Project Management is easily the most populous techniques amongst Project Managers. Developed in the 1950s, the Critical Path technique focuses its strength on the time frame within different tasks and also the level of dependencies surrounding it. Instead of just checking your progress like that of the PERT technique, this particular technique focuses more on your priorities and how to ensure you don't stay stagnant. It would make sure it gives you the right measurement and estimates of a task, thereby, equipping you with whatever you need to make the task a success.

6.   Critical Chain Project Management: This is an improved version of the Critical Path technique. Its focus on the resizing and cut-down of the project team, budget, and so much more. It believes that if that happens, the amount of pressure on the project would also reduce, thus, leading to success in the end. Instead of speeding up the process as Critical Path suggested, you follow the estimated data and use that as a means in cutting down unnecessary things surrounding the project. This is one of the most competitive techniques used in recent times.

Key Methodologies in Project Management

1.   Agile: This is no doubt one of the most populous Project Management methodologies that had ever been developed. If you are working on a project that has a high chance of been repeated, then this is the methodology for you. With Agile, you can't do it alone. You will need the maximum cooperation of your team and that of your customers (in the form of feedbacks) in order to be able to find probable solutions to the problems surrounding your project. The Agile was first created and tested for software development use only until it was used outside that line. Ever since it had been a driving force for every type of Project Managers out there.

As a direct improvement on the waterfall technique, this methodology gates it's tenets from the principles of Agile Manifesto. What is an Agile Manifesto? I will tell you. In 2001, a group of amazing and outstanding leaders of various industries (about 13 in number) made a declaration on unraveling legitimate and better methods for software development, thereby, ensuring these methods still follows a path where there would be iterative development, teamwork, and the necessary changes.

Agile is a very adaptive methodology which can be used in a whole lot of project type. It is important to know that Agile is popularly used by lots of Project Managers out there to drive home a rather difficult project. Agile also features a whole lot of functions which includes the use of the six main deliverables (product vision statement, product roadmap, product backlog, release plan, Sprint Backlog, and increment) in making sure the project remains unshaken. That way, it will further help develop an outstanding product in the end.

2. Scrum: This particular type of methodology has to deal with five different values which are commitment, courage, focus, openness, and respect. Just like Agile, it also aims at making projects reach their desired destinations. It is also perfect for the iterative type of projects. Scrum will make you deliver and develop amazing projects which will blow the minds of your clients away. Agile and Scrum are in fact two sides of the same coin. What differentiates them is the way and manner in which they are operated, else the functions and goals still remain the same.

Now, how do Scrum operate? It's quite easy. Scrum draws its strength by focusing on specially assigned roles in the team, certain events that may occur, and other artifacts surrounding the project. The roles are strictly divided into three – Product Owner, Development Team, and the Scrum Master. Scrum events are specially designed events that would make your goal become more achievable. They are the Sprint, Sprint Planning, Daily Sprint, Sprint Review, and the Sprint Retrospective,

The Scrum artifacts are in the form of logs which contains all the important list of priorities and sometimes irrelevances. This is to be double sure and not to give a chance to further problems. The Scrum artifacts include the Product Backlog and the Sprint Backlog. The former is held by the Product Owner. As a representative of the client or stakeholders, he or she has every right to hold this list. It includes everything about the project – costs, benefits, budgets, functions, and so much more. The Sprint Backlog contains the necessary information for the execution of the next sprint.

3.   Kanban: We have already made references to this in one of our earliest chapters. Believe it or not, Kanban is making waves and is a methodology that is trending in today's world. Just like Scrum, it also has the same tenets with Agile. Initially, Kanban was a brainchild of the Toyota production organizations during the 1940s and had ever since found its way into the world of Project Management. How did this happen? It's quite simple.

Kanban is a simple framework that focuses on the use of pictorial and high quality painted spread across a board in order to a team to visualize and have a clear perception of what they had accomplished, what their challenges are, and what their target should be. Kanban is all about effectiveness and efficiency in a team. If the team possesses this amazing quality, then trust me, Kanban is the best methodology you can ever think of. It has a unique way of bringing loopholes to light no matter how hard they must have hidden.

You can also use so many cues in Kanban. Cues like the Kanban Boards which be used in holding the other tools of making a Kanban. It is more like the mother to all Kanban cues. The Kanban Cards are the working hands of the whole Kanban process. Each card depicts a different meaning so as not to confuse anyone in the team. This would aid communication and make everything clear. Then the Kanban Swimlanes are what would aid you in differentiating the tasks in the visual board. Be that as it may, this is what Kanban really entails.

4.   Lean: This is another whole new discussion entirely. It would interest you to know that it has no connection with the Agile method. This methodology focuses on the end result which involves the customers and also tries to cut waste in any

way that it can. Instead of having a large pool of resources at your disposal and end up having very limited resources, this methodology is the best method you can employ in order to come out with an amazing result.

This methodology is also gotten from one of the Japanese fastest growing industry, the manufacturing industry. With Lean, wastes are reduced, the quality will improve drastically, and the cost will also be reduced. That way, it going to be a win-win situation for everybody. There are three types of Lean methodology and they are Muda, Mura, and Muri. These can also be known as the 3Ms of Project Management.

Muda seeks to eliminate waste. This is the primary function of Muda. Anywhere it sees irrelevances, it cuts it off, especially if it doesn't add any value to the project. And when we are referring to wastes, we are talking about time wasting, resources wasting and so much more. Mura also is all about elimination. But instead of focusing on wastes, it focuses its lens on the obstruction and variances present in the project. This would, in turn, give it the necessary flow it deserves for a successful Project. Then Muri is all about making sure the project stays afloat. It doesn't waste time in eliminating excess load and baggage the project is carrying.

5. Waterfall: The waterfall methodology had been in existence longer than many of the methodologies out there in the world of Project Management. It is a particular type of methodology which is designed in a downward slope. This downward slope is what will give you a waterfall shape at the end of your project design. This methodology was created by Winston W. Royce in his article of 1970.

This methodology focuses more on the documentation process. It deals with order in a project. For example, this methodology would give you the idea of making necessary arrangements where necessary. It would show you where you need to make adjustments. If one of your team members resigns, you will be able to fill in the new recruit through the documentation you've already filed out with the aid of this methodology. This type of methodology is also used in Software Project Management.

6.   Six Sigma: The Six Sigma is a type of methodology that was first used in 1986 by the Motorola Company. A group of engineers came up with this methodology in order to improve the quality of their products and also to tackle the errors that may occur. The Six Sigma would help you identify possible problems that may affect your project by pointing towards the direction of every flaw. It might also involve statistics and empiricism. Thus, you will need people that are skilled in this method.

There are different types of Six Sigma methodology and these are Six Sigma Green Belts and the Six Sigma Black Belts. These two types of Six Sigma are supervised by the Six Sigma Master Black Belts. There is also DMAIC and DMADV which are used in making sure the project goes smoothly. The Lean Six Sigma method would also help eliminate unnecessary waste.

Key Tools in Project Management.

1.   Gantt Chart: This is one of the best scheduling tools you will ever find. It is easy to use and operate. Gantt Chart is a Project Management tool which can be used in showing the progress of each task and activity in a project. Developed in

1917, the Gantt Chart shows the priorities in the tasks as well as their connections. Ever since it was introduced, it had become a driving force in today's world. The Gantt Chart was first used during the creation of America's Hoover Dam project of 1931. As a beginner in this line of professionalism, I would employ you to take this seriously as it would always come in handy.

2.   Logic Network: This is a certain network that shows the arrangements of activities and events during a specific time. That way, you will be able to foresee the next line of action in a project. This can also be used to identify the priorities in a project. The Logic Network will allow you to see things beforehand, that way, you will be able to predict the future and even tweak the project for a favorable outcome. This tool will even open your eyes to the possibilities around you. Additionally, important information would definitely surface no matter how hidden they may be.

3.   PERT Chart: The Program Evaluation and Review Technique which is also known as PERT Chart is an important tool of Project Management created by the United States Department of Defense of the US Navy Special Projects Office. It was developed and formulated as a result of the Polaris mobile submarine-launched ballistic missile project which was carried out in the year 1958. The PERT Chart is similar to the Gantt Chart. They both would tell you the exact that you can use to execute a task in a project.

4.   Product Breakdown Structure (PBS): This is a tool of Project Management that shows the total connection and relationship of the products development and deliverables. The PBS comes in a kind of hierarchical tree or structure which

seeks to point out the goals and objectives of the project. According to PBS, before developing a relationship as regards the product range, you should first make sure the structure is well grounded.

5.   Work Breakdown Structure (WBS): Like the PERT Chart, the Work Breakdown Structure (WBS) is also one of the popular tools developed and created by the United States Development of Defense (DOD) all thanks to the Polaris mobile submarine-launched ballistic missile project in the year 1958. What the WBS seeks to do is to resize the deliverables so as to enable the tasks to be minimal in cost. The WBS is basically the base of every Project Planning.

In this chapter, we delved into the key techniques, methodologies, and tools which will come in handy during your projects. Whenever you feel your project is becoming difficult or complex, these techniques will help you move past that challenge. It will help you reach the desired destination. Additionally, you can mix them up or even use them one after the other. The choice is all yours. What should matter is the end result of the project. Additionally, there is no project without risk. No matter how little they may be, a risk is a risk. If not tackled appropriately, they might end up biting you in the butt. Thus, this is what the next chapter would talk about in details. Please turn over the page.

Chapter Eleven

Common Risks and Their Control.

There is no business without risk, neither is there any form of event, activities, strategy, and management without its own share of risks. In the same vein, Project Management is no saint in this line of classification. As a matter of fact, it's amazing and outstanding features which had made it universal all over the world also comes with its own blemish. No matter how hard you turn this, these risks are still probable circumstances that may occur along the line. However, knowing about them would give you a head-start as a beginner.

When you know what to expect, we will be able to turn the tides in our favor. This is what this chapter promises to unfold. Trust me, some projects are risk carriers in hiding. If you don't do your findings and study appropriately, you might not see the impending doom coming. This always happens to beginners who want to prove themselves in the labor market. No one will tell you if a project is risky. The client would want you to deal with every issue professional. That is why he or she is appointing you in the first place. But as a certified Project Manager, you should be able to tell if a certain project is risky or not.

Some project might not even show this at first. But along the line, you will start noticing different types of challenges coming up to frustrate your plans. Be that as it may, some projects are risky at the beginning and some are as a result of our decisions and actions. This is why we really need to check every decision and action, weigh out the pros and cons, and consult the team or client before taking it. Nevertheless, this chapter would help sensitize you with the common risks you will definitely face as a Project Manager, be it an expert or beginner. Risks know no master.

Either small or big, deep or shallow and simple or complex, risks can spoil our plans and drag our project back. It can even go a long way in destroying our plans. As a beginner, if you are faced with risk in your project, then do not panic. Panic won't solve the situation for you but instead, look closely at the situation and find ways in which you can turn the situation around for the better. Every twist and turn in a project comes with its own circumstances, thus, I will advise you to be very careful before making any decision. One wrong move and it's game over. Now, here are some of the common risk you will face while taking up a project.

1.   Wrong Estimate: This happens all the time and as a beginner, it's definitely going to happen to you too. Often times, the project gets stretched as a result of a miscalculated estimate of the amount of time for each schedule, the total amount of time the project will be executed, the amount of cost it would take for the project to be executed, the actual budget of the whole project, and so much more. When this happens, then you should know that your project would face a whole lot of challenges along the line. However, it's one of the common risks that may occur and you should watch out for.

Solution: I will recommend you choose someone very good at the estimate for this task. Not just everyone in the team is meant for this particular task. Additionally, with the new trend of AI, I will advise you not to slack off. Join the moving train and stay abreast with the trend. Make use of the different applications available today as they will make your work easy and efficient. And if the mistake is already made, then be fast about the solution. Don't let it drag for long before making the necessary adjustment.

2.　　Frequent change of goals and objectives. This is one annoying type of risk you might encounter along the line of your project. For example, the client might want to have a change of heart and decides he doesn't want the project to go a certain direction any longer. They would want you to make the necessary adjustment without even looking at the cost and how much problem it would bring for you as the Project Manager. When this happens, what will you do as a Project Manager? How would you solve this problem?

Solution: Solving this is quite simple but very difficult at the same time. I know for a fact that the answers or solution to this problem lies in inculcating and infusing the necessary changes in the project but this comes with a price. For example, the project will even take more time to be completed, there would be another whole new round of calculations, every process would have to be revisited and at the same time tweaked to the desired form.

3.　Unforeseen Circumstances: These are the type of risks we won't even see coming. They are there but we might not see them at first. But along the line, they will start showing themselves. It's better to prepare for these circumstances because they might end up spoiling your hard work. For example, a crucial member of your team might decide to travel or take a leave of absence, what if you don't or can't find a replacement almost immediately? This would definitely cost you and your project a whole lot.

Solution: Always make sure you have a backup plan before taking up a project while executing it and at the end of the project. This would come in handy, trust me. Thus, there would be no problem whatsoever when these problems occur. When these unforeseen circumstances showcase themselves, then the project would definitely get delayed, therefore, always be ready for the worst. That is the only way you would be able to overcome this problem and stay afloat with your project.

4.   Risk of communication: A lot of Project Managers suffer from this particular risk all the time. I would like to stress the fact that on no circumstances should you imagine someone has gotten and understood your directives perfectly without having to repeat the directives over and over again. Remember that the project you will handle is delicate and should be handled with utmost concentration. Always make yourself clear. When they don't really understand you, they end up doing something wrong. This would affect your project sooner or later. Communication is the key to a successful Project, never forget that.

Solution: Always make sure you lay emphasis on your words when addressing your team either individually or generally. Make sure your words are crystal clear. This would reduce the rate of misunderstanding in the team. Make easy communication a priority in your team as this would make your team operate even better in the long run. Additionally, always create a mild atmosphere in the team. Be free to them and always tell them to come to you in cases of any misconception.

5.   Overlooking design: Designs are there for a reason. They are the framework of the whole project. They showcase how best you can go about executing a project. In order words, pay close attention to the designs, they might help you point out loopholes in the project. However, a lot of expert Project Managers out there ends up overlooking the project designs and go with their instincts just to save time. Trust me, this might look cool but please don't emulate them as a beginner. Many of these experts end up making a mistake in the end. In order to save time, they tend to cut this process, thus, end up destroying their hard work.

Solution: Don't even try it. You might be lucky when you try it but the odds are greatly not in your favor. The Project design is part of the project for a reason, why not focus on it? Don't neglect it for no just reason and if you must, then be sure to know what you are doing.

6.   Technical risks: This type of risks are also challenging as they come with their own level of disadvantages. For example, when the client cuts the budget of the project or decides to cut down the strength of your team, this would definitely affect your project one way or the other. You will now have to perform more functions, meet deadlines, and at the same time uphold the quality of your work.

Solution: Beforehand, I would recommend you learn how to work under unfavorable conditions. Remember, this ought to be one of your amazing superpowers as a Project Manager. You should also be able to multi-task. Thus, immerse yourself with all these attributes beforehand and work will be like play even in unfavorable conditions.

Risks may even come every day on the project. These may be tangible or mere ones. But when they come, do not panic. It's not the end of the world. Sometimes, these little ones sometimes come as a warning sign for the complex ones to come. Thus, I will advise you to stay prepared and ready to tackle them as they come. The next chapter would put the lid on our bottle of knowledge. You don't want to miss it.

Chapter Twelve

Tapping Into the realm of Possibilities

Well, there you have it, Project Management at your fingertips. Even the so-called topnotch Project Managers also began as a Beginner. However, with dedication, commitments, and perseverance, they triumph and moved up the food chain. This can be you if only you learn to start believing in yourself and also start practicing your skills in Project Management. You can now start using these management skills in your everyday life. A lot of people have this notion of Project Management being a professional skill. This is very wrong. Being a Project Manager also makes you a Life Project Management expert.

It is now up to you to start putting your life in order. With all the knowledge you have gotten from the beginning of this book, you should be able to perform the function of a Project Manager without much stress. Project Management is a very large field, thus, you should choose which area you would want to specialize in. This is better than wanting to focus on every aspect of the field. It might be to complex for you to handle. Thus, focus your lens on any of the types of Project Management in order to come out as an expert in no time.

This book is a gift, use it wisely. And should in case you've already started your Project Management career and needed a kick in the butt, then this book can be your strong foot. Trust me, it will equip you with the necessary tools, methodologies, and techniques in which you would need in everyday Project Management. These topnotch Project Managers won't be nice. Don't think they will receive you with welcome hands just because you are new in the field. They will squeeze every ounce of opportunities that will come your way. They will compete with you for every possible project out there no matter how little.

Thus, I will urge you to be prepared. Make sure you are always ready for any competition whatsoever. Additionally, try and always boost your resume or portfolio in any way that you can. This would also come in handy. A Project Manager with a very strong portfolio finds it very easy to relate with clients and the team in general. Take that Management classes, either in person or online. Make sure you stay committed to building yourself in this field. Let every day count as you continuously strive towards developing and improving yourself with new experiences and knowledge.

And when the challenges come, remember that there is no business without setbacks neither is there a project without challenges. Take them in good faith. That is the only way you can overturn the situation. In case you don't know how to overturn a certain challenge, then look for the situation inwardly. Search within yourself and look for what you are lacking as a Project Manager. Sometimes, Project Managers Sometimes, Project Managers often face the problem of not believing in their selves, thus, give up when the going gets tough. If these topnotch Project Managers had given up halfway, they wouldn't have been basking in the euphoria of success today.

Project Management is a trending line of professionalism today that will give you the career path that you seek. Aside from its amazing prospects and benefits, it comes with a particular type of joy which is derived from the organization, planning, and execution of projects. There is this particular joy that crept in when one is at the forefront of something productive. This is why many Project Managers out there perform their functions like it's nothing. When work becomes fun, the rest will become easy. If Project Management is your passion and you don't know how to go about it, then this book will help you through that phase.

In the end, you will be finally introduced into a world of Project Management. At the same time, you would be able to put your life in order and good shape. There would nothing you won't be able to handle. No matter how complex the event or activity might be, you will be able to make something meaningful out of it. Even if all the odds are against you, you will still be able to make something meaningful out of the situation. This is the goal of Project Management. It's not just about knowing how to use the experience and knowledge gotten on projects but to also apply it on everyday life. Life would definitely become bearable and meaningful.

As a Project Manager, you will be the to-go person. You will be the person everybody runs to in time of need and problem. You will be the fixer for everyone. In as much as the attention might get into your head, I will implore you not to forget the basics. Never forget to run a check or study on the project before embarking on it. The risks would always be there, don't overlook any of them. Make sure you are familiar with the kind of risk you will encounter in the course of the project. That way, you will be able to plan ahead and make necessary arrangement in tackling them when the need arises.

Also, your team isn't mannequins, they are your shadow. Make good use of them as they are there for a reason and that is to make your work easy and faster. Always consult them when you feel there is a need for that. Wherever you are not sure or clear, be free to call those vast in the area to make it clear. You are the head, thus, any wrong move you make will definitely affect the whole team and vice versa. Thus, always check with the team before deciding on an important decision. Know that each team member is a specialist in a particular aspect. Therefore, be sure to share the task according to their skills.

Go out to the world and establish yourself as a topnotch Project Manager in the making. Ripple by ripple, you will definitely attain your dream in this field of professionalism. All you need to do is to be patient and confident. Show how much you've learned and how much output you are willing to give to your clients and stakeholders. Additionally, hold your client close. You don't want to be seen as incompetent on your first project. That will reflect very badly on your resume. And remember first impression matters in this line of business. Make sure you sweep them off their feet completely at the first meeting. This would make them go soft on you.

And one thing about project management is that your clients get to call you back for a project when they feel you are capable of. Thus, the first impression is all you need. Do that and your name would be on the lips of every Project Manager in the country. It would only be a matter of time before you go international.

And that wraps it up on how to become a Project Manager. This process is no joke. It is not as easy as it sounds. In fact, it requires lots of hard work, perseverance, and skill in order for a beginner to move up the food chain in this line of professionalism. Nevertheless, I believe you will attain that position, all you need to do is just to believe in yourself and your potentials. The sky would be your starting point afterward.

Conclusion

Without mincing many words, this has been an amazing journey all through the chapters. In the end, I am certain you must have gotten the answers you are certain for. And if you still haven't gotten your answers as regards Project Management, then I will suggest you look inward. Look closely at yourself and point out how you feel. Remember, Project Management and Emotional Intelligence goes hand in hand. Go for emotional intelligence classes if you must. It would help you contain your feelings and make you feel better afterward.

You can't be the amazing Project Manager you had always dreamed of if you don't get your act together. First, you need to have a clear mind to make that happen. Be that as it may, go out there and make history. Go out there and make a name for yourself. After all, this is why you have been focused, committed, and patient.

Now, let's do a quick recap, shall we? In case you had forgotten what our chapters looked like, chapter one delved into the historical background of Project Management. Questions like how did Project Management come into being? Who and why was Project Management coined? And so much more were answered in this chapter. Chapter two of this book also defined Project Management along with the key concepts associated with it. We can go into Project Management without taking a walk into this line of thought. The chapter highlighted scholarly definitions as well as that of the concepts.

Chapter three focuses its lens on how one can become a Project Manager. This is one of the reasons why a lot of you had picked up this book. Make sure you pay close attention to the chapter as it explains everything in general. Chapter four also looked at the duties, roles, and functions of a Project Manager. The role of a Project Manager goes beyond just managing. This chapter shows you that as it sheds more light on what it takes to be a real Project Manager. Chapter five will show you tips on how to be successful in this field of professionalism. You might want to start infusing to your everyday endeavors.

Chapter six looked at the common mistakes you might make as a beginner. Mistakes have no master, thus, don't feel bad as even the best in this field make mistakes too. Chapter seven also holds something interesting. It will sensitize you on the new trends and also focus on the kinds of trends we have in today's world. You don't want to be left out. Chapter eight also delved into Project plan and to go about making an amazing plan for the success of your project. If I were you, I will utilize this gift appropriately. Chapter nine contains Project Control. If you don't have this, then you are not fit to be a Project Manager. This chapter would teach you how to gain control of your project.

And chapter ten, eleven, and twelve looks at the key techniques, methodologies, & tools, common risks, and their control, and tapping into the realm of possibilities respectively. These chapters would further broaden your horizon in more ways you can ever think of. With that being said, it's left for you to make use of what you had assimilated so far. All I can tell you now is good luck and thank you for sticking with us this far.

All the best and God bless!

www.ingramcontent.com/pod-product-compliance
Lightning Source LLC
Chambersburg PA
CBHW071202210326
41597CB00016B/1641